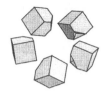

Origami³

Third International Meeting
of Origami Science, Mathematics, and Education
Sponsored by OrigamiUSA

Thomas Hull, Editor

A K Peters
Natick, Massachusetts

Editorial, Sales, and Customer Service Office

A K Peters, Ltd.
63 South Avenue
Natick, MA 01760
www.akpeters.com

Library of Congress Cataloging-in-Publication Data

International Meeting of Origami Science, Math, and Education (3rd : 2001 : Asilomar, Calif.)
    Origami3 : Third International Meeting of Origami Science, Math, and Education /
Thomas Hull, editor.
        p. cm.
    Includes index.
    ISBN 1-56881-181-0 (pbk.)
    1. Origami--Mathematics--Congresses. 2. Origami in education--Congresses. I. Title:
Origami3. II. Title: Third: Third International Meeting of Origami Science, Math, and Education.
III. Hull, Thomas. IV. Title.

    QA491 .I55 2001
    736í.982--dc21

                                                                                    2002023871

Printed in Canada
06 05 04 03 02   10 9 8 7 6 5 4 3 2 1

Table of Contents

Part Two: Origami Science and Applications   119

Part Three: Origami in Education   233

Preface

It is with much excitement and pleasure that I write the preface to *Origami³*, *Proceedings of the Third International Meeting of Origami Science, Math, and Education (3OSME)*. As program organizer of this conference and editor of this proceedings, this book symbolizes the completetion of three years of work.

The first such conference, held in Ferrara, Italy in December of 1989, was titled *The First International Meeting of Origami Science and Technology*. Humiaki Huzita of the University of Padova was the father of this historic event. There were only 17 talks, but the foundation was laid for people around the globe to share their interests and research on the scientific, mathematical, and educational aspects of origami. The proceedings from the first meeting were ground-breaking, being the first publication to collect and disseminate paper-folding research. To this day, these proceedings are much sought-after.

The Second International Meeting of Origami Science and Scientific Origami was held in Otsu, Japan in December of 1994; this was the first such meeting that I was able to attend. I was very impressed with the variety and scope of the talks, which covered everything from Miura's uses of origami in designing solar panel arrays to Yoshizawa's views on origami art; from Kawasaki's work on abstract, high-dimensional flat-folding to Kresling's studies of origami patterns in nature. The juxtaposition of ideas in science, art, education, and math made the Otsu meeting unique in the history of origami. This meeting also gave birth to a wonderful proceedings book; the organizers of the meeting and editors of the proceedings (Miura, Fuse, Kawasaki, and Maekawa) should be commended for their outstanding achievement.

During the 1994 Otsu meeting, it was suggested that the third such meeting should take place in the United States. The next year, I was in San Francisco for a conference and was fortunate enough to have dinner with Peter Engel (author of *Folding the Universe*) and Robert J. Lang. When I told them about the Otsu conference I had been to and how the next meeting should be in the USA, they replied that I should be the one to organize it. I was in graduate school at the time, and couldn't conceive of how I was going to organize such an event. Now that it has happened, I must confess that it feels like I have fulfilled one of the purposes of my life.

Of course, OrigamiUSA was the natural choice to sponsor and organize the event. As an OrigamiUSA board member, I chose to take the lead in this project, and they provided all the logistical help in putting the conference together. One of OrigamiUSA's missions is to disseminate the knowledge and educational uses of origami to the public, and through this conference, they certainly furthered that goal. For more information about OrigamiUSA, or to learn about becoming a member, please write to them at OrigamiUSA, 15 West 77th Street, New York, NY 10024-5192 or visit their web site at: http://www.origami-usa.org.

We are also very fortunate to have found a publisher for these proceedings, A K Peters. Through their means as a publisher, these proceedings will find their way into the libraries of the world and become known to mathematicians, scientists, and educators. For the first time, scholarly work on origami will be easily available for those seeking it. I, for one, am very glad that I will now have somewhere to direct all the people who ask me every year, "Where can I learn more about origami and math/science/education?"

I also had personal reasons for wanting to make the third meeting as successful as possible. Some of the first research I encountered on the mathematics of origami came from the first proceedings book of the 1989 meeting. It is not an exaggeration to say that reading these articles while I was in graduate school studying to be a mathematician changed my life. They opened my eyes to the vast potential for serious work in the study of paperfolding. Since then, my career path and life's dream has been to pursue study, both in math and in education. It is my deep hope that through 3OSME, and through the proceedings you now hold in your hands, we will extend the tradition of exploring the joy of origami knowledge in all fields of study and continue the wonderful efforts of our predecessors.

A number of people were instrumental to the development of these proceedings. sarah-marie belcastro, V'Ann Cornelius, and Robert Lang provided much-needed editorial help and advice on a number of the papers. Anne LaVin was an absolute godsend in helping the editor deal with the multitude of graphic images that were sent to the author in a variety of international formats.

Alice and Klaus Peters (the "A" and "K" of A K Peters) deserve a lot of credit for further editorial help and the wonderful final layout of the book. Finally, all the contributing authors need to be thanked (including those whose work was not able to be included) for allowing the 3OSME conference and these proceedings to be of such high quality.

Thomas C. Hull
3OSME Editor
Department of Mathematics
Merrimack College

Part One

The Mathematics of Origami

By "the mathematics of paper-folding," we mean an attempt to learn the "hows and whys" of paper-folding by modeling the process with mathematics. To most people who have actually done some origami, this should seem very natural. Anyone who folds, say, the traditional crane will probably, at some point, unfold one and marvel at the intricate lattice of crease lines it reveals. At such moments, the link between geometry and origami seems crystal-clear, but rules governing such links appear quite nebulous. Explaining this mystery is the goal of origami mathematics, as well as of the authors of the papers in this section.

It has been known since the 1930s that simple paper-folding is more powerful as a geometric construction device than the classic straight-edge and compass. Much remains to be understood about how far origami constructions can be pushed. Alperin, Geretschläger, and Scimemi provide us with further advancements in this area.

An axiomatic, construction approach to modeling origami does not capture the whole picture, however. Indeed, trying to devise a more general mathematical setting in which to study paper-folding is not an easy task. Belcastro and I describe how both flat and three-dimensional origami can be explored using combinatorics and linear algebra.

A more recent development in the study of origami math has been to delve into questions of computational complexity: How long would it take a computer to decide if a given crease pattern will fold flat, or if a certain shape can be folded? Demaine and Demaine give a broad and detailed survey of this field, while Bern, Demaine, Eppstein, and Hayes solve a specific problem of this type that leads to a fun origami magic trick.

Most research on modular origami has been geared toward using geometry to develop new and more intricate origami designs. (See the papers by Lang and the Burczyks in the other sections of this proceedings for examples of such work.) This is different from trying to explore the mathematical processes that govern how modular origami units fit together to form objects. Mosely presents one of the first treatments of such geometry at work in modular origami models. The motivation is to understand which modular origami models are actually valid polyhedra.

Getting back to basics, the mathematics of even simple origami models can reveal a lot of unexplored ground. Kawasaki explains in detail the flexible geometry in folding the classic Japanese crane, or *ori-zuru*, while Maekawa gives the first complete mathematical definition of iso-area folds, a beautifully symmetric class of origami models.

At the 3OSME conference, we were very fortunate to have such a fascinating and varied list of speakers on origami mathematics, and we are very grateful for their efforts to provide such an interesting array of papers for this proceedings.

Thomas C. Hull

Recent Results in Computational Origami

Erik D. Demaine and Martin L. Demaine

1 Overview

Computational origami is a recent branch of computer science studying efficient algorithms for solving paper-folding problems. Most results in computational origami fit into at least one of three categories: universality results, efficient decision algorithms, and computational intractability results.

A *universality result* shows that, subject to a certain model of folding, everything is possible. For example, any tree-shaped origami base (Section 2.1), any polygonal silhouette (Section 2.3), and any polyhedral surface (Section 2.3) can be folded out of a large enough piece of paper. Universality results often come with efficient algorithms for finding the foldings; pure existence results are rare.

When universality results are impossible (some objects cannot be folded), the next-best result is an *efficient decision algorithm* to determine whether a given object is foldable. Here, "efficient" normally means "polynomial time." For example, there is a polynomial-time algorithm to decide whether a "map" (grid of creases marked mountain and valley) can be folded by a sequence of simple folds (Section 3.4).

Not all paper-folding problems have efficient algorithms, and this can be proved by a *computational intractability result*. For example, it is NP-hard to tell whether a given crease pattern folds into any flat origami (Section 3.2), even when folds are restricted to simple folds (Section 3.4). These results mean

3

that there are no polynomial-time algorithms for these problems, unless some of the hardest computational problems can also be solved in polynomial time, which is generally deemed unlikely.

We further distinguish computational origami results as addressing either *origami design* or *origami foldability*. Basically, in origami design, some aspects of the target configuration are specified, and the goal is to design a suitable target that can be folded out of paper. In origami foldability, the target configuration is unspecified and arbitrary, and rather the initial configuration is specified, specifically the crease pattern possibly marked with mountains and valleys, and the goal is to fold something (anything) using precisely those creases. While at first, it may seem that an understanding of origami foldability is a necessary component to origami design, the results indicate that, in fact, origami design is much easier to solve than origami foldability which is usually intractable.

Our survey of computational origami is partitioned accordingly into Section 2 (origami design) and Section 3 (origami foldability). In addition, Section 4 briefly surveys some of the other problems in a more general field of geometry called *folding and unfolding*. Specifically, we describe a one-dimensional, linkage-folding problem (Section 4.1) and unfolding polyhedra into nonoverlapping nets (Section 4.2).

2 Origami Design

We define *origami design* loosely as, given a piece of paper, fold it into an object with certain desired properties, e.g., a particular shape. Most closely related to "traditional" origami design is Lang's TreeMaker work (Section 2.1), which has brought modern origami design to a new level of complexity. Related to this work is the problem of folding a piece of paper to align a prescribed graph (Section 2.2), which can be used for a magic trick involving folding and one complete straight cut. Another approach is to design an origami with a specific silhouette or three-dimensional shape (Section 2.3), although the algorithms developed so far do not lead to practical foldings. A recent specific type of origami is an origami tessellation (Section 2.4), which "folds" a tiling of the plane. A nonstandard form of origami is to start with a piece of paper that is not flat, but rather the surface of a polyhedron, and the goal is to flatten that surface (Section 2.5).

2.1 TreeMaker

TreeMaker is a computer program by Robert Lang that implements the *tree method* for origami design. Some components of this method, such as special cases of the constituent molecules and the idea of disk packing, as well as other methods for origami design, have been explored in the Japanese technical origami community, in particular by Jun Maekawa, Fumiaki Kawahata, and Toshiyuki Meguro. This work has led to several successful designs, but a full survey is beyond the scope of this paper; see [24]. Here we concentrate on Lang's work [23, 24]; over the past several years, he has developed the tree method to the point where an algorithm and computer program have been explicitly defined and implemented.

The tree method allows one to design an origami *base* in the shape of a specified tree with desired edge lengths, which can then be folded and shaped into an origami model. More precisely, the tree method designs a *uniaxial base* [23], which must have the following properties: The base lies above the xy-plane, all faces of paper are perpendicular to the xy-plane, the projection of the base to the xy-plane is precisely where the base comes in contact with the xy-plane, and this projection is a tree. The theorem is that every geometric tree (an unrooted tree with cyclicly ordered children and prescribed edge lengths) is the projection of a uniaxial base, e.g., folded from a square.

The crease pattern for such a base can be found efficiently, certainly in $O(N \operatorname{polylog} N)$ time where N is the number of creases (the output size). It is unclear whether the number of creases required is bounded in terms of the combinatorial complexity of the input, i.e., the number of vertices in the input tree. Optimizing the base to make maximal use of the paper is a difficult nonlinear constrained optimization problem, but the TreeMaker software has shown the viability of existing methods for finding good local optima. Indeed, additional practical constraints can be imposed, such as symmetry in the crease pattern, or that angles of creases are integer multiples of some value (e.g., 22.5°) subject to some flexibility in the edge lengths.

TreeMaker finds a crease pattern that results in the desired uniaxial base. Work on a related problem (see the next section) by Bern, Demaine, Eppstein, and Hayes [4] suggests a method for finding an appropriate mountain-valley assignment for the crease pattern, and possibly also the resulting folded state.

2.2 One Complete Straight Cut

Take a piece of paper, fold it flat, make one complete straight cut, and unfold the pieces. What shapes can result? This *fold-and-cut* problem was first formally stated by Martin Gardner in 1960 [16], but has a much longer history, going as far back as 1721 (see [9]).

More formally, given a planar graph drawn with straight edges on a piece of paper, can the paper be folded flat so as to map the entire graph to a common line, and map nothing else to that line? The surprising answer is

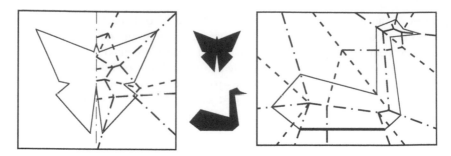

Figure 1. Crease patterns for folding a rectangle of paper flat so that one complete straight cut makes a butterfly (left) or a swan (right), based on [9, 10].

that this is always possible, for any collection of line segments in the plane, forming nonconvex polygons, adjoining polygons, nested polygons, etc. There are two solutions to the problem. The first solution [9, 10] is based on a structure called the straight skeleton, which captures the symmetries of the graph, thereby exploiting a more global structure to the problem. The second solution [4] is based on disk packing to make the problem more local, and achieves efficient bounds on the number of creases. See Figure 1 for examples of the first method, and [4] included in this proceedings for an example and more information on the second method.

While this problem may not seem directly connected to pure folding because of the one cut, the equivalent problem of folding to line up a collection of edges is, in fact, closely connected to origami design. Specifically, one subproblem that arises in TreeMaker is that the piece of paper is decomposed into convex polygons, and the paper must be folded flat so as to line up all the edges of the convex polygons, and place the interior of these polygons above this line. The fold-and-cut problem is a generalization of this situation to arbitrary graphs: nonconvex polygons, nested polygons, etc. In TreeMaker, there are important additional constraints in how the edges can be lined up, called path constraints, which are necessary to enforce the desired geometric tree. These constraints lead to additional components in the solution called *gussets*.

2.3   Silhouettes and Polyhedra

A more direct approach to origami design is to impose the exact final shape that the paper should take. More precisely, suppose we specify a particular flat silhouette, or a three-dimensional polyhedral surface, and desire a folding of a sufficiently large square of paper into precisely this object, allowing coverage by multiple layers of paper. For what polyhedral shapes (shapes made up of flat sides) is this possible? This problem is implicit throughout origami design, and was first formally posed by Bern and Hayes in 1996 [5]. The surprising answer is "always," as established by Demaine, Demaine, and Mitchell in 1999 [13].

The basic idea of the approach is to fold the piece of paper into a thin strip, and then wrap this strip around the desired shape. This wrapping can be done particularly efficiently using methods in computational geometry. Specifically, three algorithms are described in [13] for this process. One algorithm maximizes paper usage; the amount of paper required can be made arbitrarily close to the surface area of the shape, which is optimal. Another algorithm maximizes the width of the strip subject to some constraints. A third algorithm places the visible *seams* of the paper in any desired pattern forming a decomposition of the sides into convex polygons. In particular, the number and total length of seams can be optimized in polynomial time in most cases [13].

All of these algorithms allow an additional twist: The paper may be colored differently on both sides, and the shape may be two-colored according to which side should be showing. In principle, this allows the design of two-color models similar to the models in Montroll's *Origami Inside-Out* [28]. An example is shown in Figure 2.

Of course, because of the reliance on thin strips, none of these methods lead to practical foldings, except for small examples or when the initial piece of paper is a thin strip. Nonetheless, the universality results of [13] open the door to many new problems. For example, how small a square can be folded into a desired object, e.g., a $k \times k$ chessboard? This optimization problem remains open even in this special case, as do many other problems about finding efficient foldings of silhouettes, two-color patterns, and polyhedra.

Figure 2. A flat-folding of a square of paper, black on one side and white on the other, designed by John Montroll [27, pp. 94–103].

2.4   Origami Tessellations

Roughly, an origami tessellation is a flat-folding of a piece of paper based on a *tessellation* or *tiling* of the plane [17]. One way to make this notion more precise [22, 35] is to consider the whole plane as the piece of paper and define a *symmetric origami tessellation* to be a flat-folding of the plane whose symmetry group is one of the 17 crystallographic groups. More generally, an origami tessellation might be defined to be any flat-folding of the infinite plane in which no bounded region of the plane contains all of the vertices of the crease pattern. In such a folding, the crease pattern (at least) will necessarily form some kind of tiling.

An origami tessellation can be related to a tiling in various ways. Typically, there are large faces in the flat-folded state that remain uncreased, and these faces are precisely the tiles of a tiling. One method for specifying creases around these faces is based on shrinking the tiles and introducing the dual tiling in the resulting gaps, using the notion of a hinged primal-dual tiling (see [36]). Bateman [3] has formalized this method to the point of a computer implementation, called Tess. See [3] for more details.

Some of the key people working on origami tessellations include Alex Bateman [3], Paulo Barreto [2, 30], S. Fujimoto [15], Thomas Hull (unpublished), Toshikazu Kawasaki [22], Robert Lang (unpublished), Chris Palmer [30, 31], and Helena Verrill [35]. Unfortunately, much of the work on origami tessellations has not been written formally, so the exact computational results are unclear. Certainly a wide range of origami tessellations have been designed, many using precise algorithms. One of the simplest class of examples are the 11 Archimedian tilings. Still open to debate is a characterization of which tilings lead to origami tessellations.

2.5   Flattening Polyhedra

In most forms of computational origami, the piece of paper is flat: a polygon in the plane, possibly with holes. Suppose, instead, we start with a piece of paper that is a polyhedral surface in space. How can polyhedra be folded? Specif-

Figure 3. Flattening a tetrahedron, from left to right. Note that the faces are not flat in the middle picture.

ically, a natural question [11] is whether every polyhedron can be *flattened*: folded into a flat origami. Demaine, Demaine, and Lubiw [11] have shown that there are flattened states of several classes of polyhedra, including convex polyhedra and orthogonal polyhedra. See Figure 3 for an example. Recently (March 2001), together with Barry Hayes, it has been shown that all polyhedra have flattened states.

3   Origami Foldability

We distinguish *origami foldability* from origami design as starting from a given crease pattern, and the goal is to fold an origami that uses precisely these creases. (Arguably, this is a special case of our generic definition of origami design, but we find it a useful distinction.) The most common case studied is that the resulting origami should be flat, i.e., lie in a plane.

3.1   Local Foldability

For crease patterns with a single vertex, it is relatively easy to characterize flat foldability. Without specified crease directions, a single-vertex crease pattern is flat-foldable precisely if the alternate angles around the vertex sum to 180°. This is known as Kawasaki's Theorem [5, 18, 20, 21]. When the angle condition is satisfied, a characterization of valid mountain-valley assignments and flat-foldings can be found in linear time [5, 20], using Maekawa's Theorem [5, 18, 20] and another theorem of Kawasaki [5, 18, 21] about constraints on mountains and valleys. In particular, Hull has shown that the number of distinct mountain-valley assignments of a vertex can be computed in linear time [19].

A crease pattern is called *locally foldable* if there is a mountain-valley assignment so that each vertex locally folds flat, i.e., a small disk around each vertex folds flat. Testing local foldability is nontrivial because each vertex has flexibility in its assignment, and these assignments must be chosen consistently: No crease should be assigned both mountain and valley by the two incident vertices. Bern and Hayes [5] proved that consistency can be resolved efficiently when it is possible: Local foldability can be tested in linear time.

3.2   Existence of Folded States

Given a crease pattern, does it have a flat-folded state? Bern and Hayes [5] have proved that this decision problem is NP-hard, and thus computationally intractable. Because local foldability is easy to test, the only difficult part is global foldability, or more precisely, computing a valid *overlap order* of the crease faces that fold to a common portion of the plane. Indeed, Bern and Hayes

[5] prove that, given a crease pattern and a mountain-valley assignment that definitely folds flat, finding the overlap order of a flat-folded state is NP-hard.

3.3 Equivalence to Continuous Folding Process

In the previous section and Section 2.5, we have alluded to the difference between two models of folding: the final folded state (specified by a crease pattern, mountain-valley or angle assignment, and overlap order) and a continuous motion to bring the paper to that folded state. Basically all results, in particular those described so far, have focused on the former model: proving that a folded state exists with the desired properties. Intuitively, by appropriately flexing the paper, any folded state can be reached by a continuous motion, so the two models should be equivalent. Only recently has this been proved formally, by Demaine and Mitchell [14], and so far only for rectangular pieces of paper.

The only other paper of which we are aware that proves the existence of continuous folding processes is [8]. This paper proves that every convex polygon can be folded into a uniaxial base via Lang's universal molecule [24] without gussets. Furthermore, unlike [14], no additional creases are introduced during the motion, and each crease face remains flat. This result can be used to animate the folding process.

3.4 Map-Folding: Sequence of Simple Folds

In contrast to the complex origami folds arising from reaching folded states [8, 14], we can consider the less complex model of simple folds. A *simple fold* (or *book fold*) is a fold by $\pm 180°$ along a single line. Examples are shown in Figure 4. This model is closely related to "pureland origami" introduced by Smith [32, 33].

We can ask the same foldability questions for a sequence of simple folds. Given a crease pattern, can it be folded flat via a sequence of simple folds? What if a particular mountain-valley assignment is imposed?

An interesting special case of these problems is *map-folding* (see Figure 4): Given a rectangle of paper with horizontal and vertical creases, each marked mountain or valley, can it be folded flat via a sequence of simple folds? Traditionally, map-folding has been studied from a combinatorial point of view; see [26]. Arkin *et al.* [1] have shown that deciding foldability of a map by simple folds can be solved in polynomial time. If the simple folds are required to fold all layers at once, the running time is at most $O(n \log n)$, and otherwise the running time is linear.

Surprisingly, slight generalizations of map-folding are (weakly) NP-complete [1]. Deciding whether a rectangle with horizontal, vertical, and diagonal ($\pm 45°$)

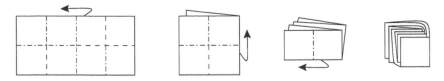

Figure 4. Folding a 2 × 4 map via a sequence of three simple folds.

creases can be folded via a sequence of simple folds is NP-complete. Alternatively, if the piece of paper is more general, a polygon with horizontal and vertical sides, and the creases are only horizontal and vertical, the same problem becomes NP-complete.

These hardness results are *weak* in the sense that they leave open the existence of a *pseudopolynomial-time* algorithm, whose running time is polynomial in the total length of creases. Another intriguing open problem, posed by Jack Edmonds, is the complexity of deciding whether a map has a flat-folded state.

4   Related Problems in Folding and Unfolding

Folding and unfolding is an area of geometry in which, like computational origami, there have been several new results in the past few years. In general, the problem of interest is how a particular object (e.g., a piece of paper, a linkage, or a polyhedron) can be folded subject to some natural constraints. For general surveys on folding and unfolding, see [7, 29].

4.1   Linkage Folding: One-Dimensional Origami

Consider a one-dimensional piece of paper—a line segment—marked at certain points with creases. It is most natural to fold such a piece of paper in the plane; a flat-folding ends with the paper back on a line. Every one-dimensional piece of paper has a flat-folded state, by alternating the creases between mountain and valley. If the mountain-valley assignment is prescribed, the crease pattern may not be flat-foldable, but this can be tested in linear time [1]. Indeed, it is proved in [1] that, in one dimension, the existence of a flat-folded state is equivalent to foldability via a sequence of simple folds.

Now consider the problem of reaching a folded state, not necessarily flat, by a continuous motion. We think of the creases (points) as *hinges*, and the edges between the creases as rigid *bars*. Can we continuously fold the one-dimensional piece of paper into a desired folded state while remaining in the plane and without introducing new creases? This carpenter's rule problem is usually stated in the (equivalent) reverse direction: Can every polygonal

Figure 5. Convexifying the "doubled tree" on the left. For more animations, see: http://db.uwaterloo.ca/~eddemain/linkage/.

chain be straightened out without self-intersection and while preserving the bar lengths? Connelly, Demaine, and Rote [6] recently proved that the answer is always yes. In addition, every polygon (polygonal loop) can be folded into a convex configuration. Some examples of these motions are shown in Figure 5. Streinu [34] has developed another motion based on [6].

4.2 Folding and Unfolding Polyhedra

A standard method for building a model of a polyhedron is to cut out a flat *net* or *unfolding*, fold it up, and glue the edges together so as to make precisely the desired surface. Given the polyhedron of interest, a natural problem is to find a suitable unfolding. For example, can every convex polyhedron be cut along its edges and unfolded into one flat piece without overlap? This classic problem, dating back to 1525, remains open. However, if cuts are allowed interior to the faces of the polygon, there is a one-piece unfolding without overlap. See [7, 29] for details.

On the other hand, given a polygonal piece of paper, we might ask whether it can be folded and its edges can be glued together so as to form a convex polyhedron. This problem has been studied by Lubiw and O'Rourke [25], and by Demaine, Demaine, Lubiw, and O'Rourke [12]. A particularly surprising

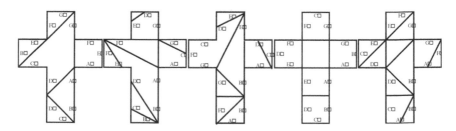

Figure 6. The five edge-to-edge gluings of the cross [25]. For a video animation of the foldings see: http://db.uwaterloo.ca/~eddemain/metamorphosis/.

discovery from this work [25] is that the well-known cross unfolding of the cube can be folded into exactly five convex polyhedra by edge-to-edge gluing: a doubly covered (flat) quadrangle, an (irregular) tetrahedron, a pentahedron, the cube, and an (irregular) octahedron. See Figure 6 for crease patterns and gluing instructions.

References

[1] Esther M. Arkin, Michael A. Bender, Erik D. Demaine, Martin L. Demaine, Joseph S. B. Mitchell, Saurabh Sethia, and Steven S. Skiena. When can you fold a map? In *Proceedings of the 7th International Workshop on Algorithms and Data Structures*, Providence, Rhode Island, August 2001.

[2] Paulo Taborda Barreto. Lines meeting on a surface: The "Mars" paperfolding. In Koryo Miura, editor, *Proceedings of the 2nd International Meeting of Origami Science and Scientific Origami*, pages 323–331, Otsu, Japan, November–December 1994.

[3] Alex Bateman. Computational tools for origami tessellations. In *Proceedings of the 3rd International Meeting of Origami Science, Math, and Education*, Monterey, California, March 2001.

[4] Marshall Bern, Erik Demaine, David Eppstein, and Barry Hayes. A disk-packing algorithm for an origami magic trick. In *Proceedings of the 3rd International Meeting of Origami Science, Math, and Education*, Monterey, California, March 2001. Improvement of version appearing in *Proceedings of the International Conference on Fun with Algorithms*, Isola d'Elba, Italy, June 1998, pages 32–42.

[5] Marshall Bern and Barry Hayes. The complexity of flat origami. In *Proceedings of the 7th Annual ACM-SIAM Symposium on Discrete Algorithms*, pages 175–183, Atlanta, January 1996.

[6] Robert Connelly, Erik D. Demaine, and Günter Rote. Straightening polygonal arcs and convexifying polygonal cycles. In *Proceedings of the 41st Annual Symposium on Foundations of Computer Science*, pages 432–442, Redondo Beach, California, November 2000.

[7] Erik D. Demaine. Folding and unfolding linkages, paper, and polyhedra. In *Proceedings of the Japan Conference on Discrete and Computational Geometry*, Lecture Notes in Computer Science, Tokyo, Japan, November 2000.

[8] Erik D. Demaine and Martin L. Demaine. Computing extreme origami bases. Technical Report CS-97-22, Department of Computer Science, University of Waterloo, Waterloo, Ontario, Canada, May 1997.

[9] Erik D. Demaine, Martin L. Demaine, and Anna Lubiw. Folding and cutting paper. In J. Akiyama, M. Kano, and M. Urabe, editors, *Revised Papers from the Japan Conference on Discrete and Computational Geometry*, Volume 1763 of *Lecture Notes in Computer Science*, pages 104–117, Tokyo, Japan, December 1998.

[10] Erik D. Demaine, Martin L. Demaine, and Anna Lubiw. Folding and one straight cut suffice. In *Proceedings of the 10th Annual ACM-SIAM Symposium on Discrete Algorithms*, pages 891–892, Baltimore, MD, January 1999.

[11] Erik D. Demaine, Martin L. Demaine, and Anna Lubiw. Flattening polyhedra. Manuscript, 2000.

[12] Erik D. Demaine, Martin L. Demaine, Anna Lubiw, and Joseph O'Rourke. Enumerating foldings and unfoldings between polygons and polytopes. In *Proceedings of the Japan Conference on Discrete and Computational Geometry*, Lecture Notes in Computer Science, Tokyo, Japan, November 2000.

[13] Erik D. Demaine, Martin L. Demaine, and Joseph S. B. Mitchell. Folding flat silhouettes and wrapping polyhedral packages: New results in computational origami. *Computational Geometry: Theory and Applications*, 16(1):3–21, 2000.

[14] Erik D. Demaine and Joseph S. B. Mitchell. Reaching folded states of a rectangular piece of paper. Manuscript, May 2001.

[15] S. Fujimoto. *Sojo suru origami asobi no shotai* (Invitation to creative origami playing). Asahi Culture Center, 1982.

[16] Martin Gardner. Paper cutting. In *New Mathematical Diversions (Revised Edition)*, Spectrum Series, chapter 5, pages 58–69. The Mathematical Association of America, Washington, D.C., 1995.

[17] Branko Grünbaum and G. C. Shephard. *Tilings and patterns*. W. H. Freeman, 1987.

[18] Thomas Hull. On the mathematics of flat origamis. *Congressum Numerantium*, 100:215–224, 1994.

[19] Thomas Hull. Counting mountain-valley assignments for flat folds. *Ars Combinatoria*, 2002. To appear.

[20] Jacques Justin. Towards a mathematical theory of origami. In Koryo Miura, editor, *Proceedings of the 2nd International Meeting of Origami Science and Scientific Origami*, pages 15–29, Otsu, Japan, November–December 1994.

[21] Toshikazu Kawasaki. On the relation between mountain-creases and valley-creases of a flat origami. In H. Huzita, editor, *Proceedings of the 1st International Meeting of Origami Science and Technology*, pages 229–237, Ferrara, Italy, December

1989. An unabridged Japanese version appeared in *Sasebo College of Technology Report*, 27:153–157, 1990.

[22] Toshikazu Kawasaki and Masaaki Yoshida. Crystallographic flat origamis. In H. Huzita, editor, *Proceedings of the 1st International Meeting of Origami Science and Technology*, pages 223–227, Ferrara, Italy, December 1989. An unabridged Japanese version appeared in *Sasebo College of Technology Report*, 24:101–109, 1987.

[23] Robert J. Lang. A computational algorithm for origami design. In *Proceedings of the 12th Annual ACM Symposium on Computational Geometry*, pages 98–105, Philadelphia, PA, May 1996.

[24] Robert J. Lang. *TreeMaker 4.0: A Program for Origami Design*, 1998. http://origami.kvi.nl/programs/TreeMaker/trmkr40.pdf .

[25] Anna Lubiw and Joseph O'Rourke. When can a polygon fold to a polytope? Technical Report 048, Smith College, June 1996.

[26] W. F. Lunnon. Multi-dimensional map-folding. *The Computer Journal*,14(1): 75–80, February 1971.

[27] John Montroll. *African Animals in Origami*. Dover Publications, 1991.

[28] John Montroll. *Origami Inside-Out*. Dover Publications, 1993.

[29] Joseph O'Rourke. Folding and unfolding in computational geometry. In *Revised Papers from the Japan Conference on Discrete and Computational Geometry*, Volume 1763 of *Lecture Notes in Computer Science*, pages 258–266, Tokyo, Japan, December 1998.

[30] Chris K. Palmer. *Bay Area Rapid Folders Newsletter*. Issues January 1994, March 1994, Winter 1995, Winter 1996, Spring 1996, Summer 1996, and Spring 1997. The last article is joint with Paulo Barreto.

[31] Chris K. Palmer. Extruding and tesselating polygons from a plane. In Koryo Miura, editor, *Proceedings of the 2nd International Meeting of Origami Science and Scientific Origami*, pages 323–331, Otsu, Japan, November–December 1994.

[32] John S. Smith. Origami profiles. *British Origami*, 58, June 1976.

[33] John S. Smith. *Pureland Origami 1, 2, and 3*. British Origami Society. Booklets 14, 29, and 43, 1980, 1988, and 1993.

[34] Ileana Streinu. A combinatorial approach to planar non-colliding robot arm motion planning. In *Proceedings of the 41st Annual Symposium on Foundations of Computer Science*, pages 443–453, Redondo Beach, California, November 2000.

[35] Helena Verrill. Origami tessellations. In R. Sarhangi, editor, *Conference Proceedings of Bridges: Mathematical Connections in Art, Music, and Science*, pages 55–68, 1998.

[36] David Wells. Hinged tessellations. In *The Penguin Dictionary of Curious and Interesting Geometry*, pages 101–103. Penguin Books, 1991.

A Disk-Packing Algorithm for an Origami Magic Trick

Marshall Bern, Erik Demaine, David Eppstein, and Barry Hayes

1 Introduction

The great Harry Houdini was one of the first to perform the following magic trick: Fold a sheet of paper so that a single straight cut produces a cut-out of a rabbit, a dog, or whatever else one likes. Whereas Houdini only published a method for a five-pointed star [11] (a method probably known to Betsy Ross [15]), Martin Gardner [10] posed the question of cutting out more complex shapes. Demaine and Demaine [5] stated this question more formally: Given a polygon with holes P (possibly with more than one connected component) and a rectangle R large enough to contain P, find a "flat-folding" of R such that the cross-section of the folding with a perpendicular plane is the boundary of P. More intuitively, a single straight cut of the flat-folding produces something that unfolds to P. A flat-folding [4, 12] is a mathematical notion, abstracting folded paper to a nonstretchable, non-self-penetrating, zero-thickness, piecewise-linear surface in $\mathbb{R}^3$.

Demaine *et al.* [6, 7] have proposed a solution to this *cut-out problem*, based on propagating paths of folds out to the boundary of the rectangle R. Here we give a more "local" solution, based on disk-packing. Our strategy is to pack disks on R so that disk centers induce a mixed triangulation/quadrangulation respecting the boundary of polygon P. We fold each triangle or quadrilateral interior (exterior) to P upwards (respectively, downwards) from the plane of the paper, taking care that neighboring polygons agree on crease orientations. A cut through the plane of the paper now separates interior from exterior.

Disk-packing has previously been used to compute triangulations [1] and quadrangulations [3] with special properties. Disk-packing, or more precisely disk-placement, has also been applied to origami design, most notably by Lang [13]. In fact, the result in this paper is in some sense a fusion of a quadrangulation algorithm from Bern and Eppstein [3] with an origami design algorithm from Lang [13].

2   Disk-Packing

Let P be a polygon with holes, strictly contained in a rectangle R. We think of P as boundary along with interior. Let PR denote the planar straight line graph that is the union of the boundary of P and the boundary of R. In this section, we sketch how to pack disks such that each edge of PR is a union of radii of disks, and such that the disks induce a partition of R into triangles and quadrilaterals. Our solution is closely related to some mesh generation algorithms [1, 3].

The disk packing starts with interior-disjoint disks. We call a connected portion of R minus the disks a *gap*. We call a gap bounded by three arcs a *3-gap* and one bounded by four arcs a *4-gap*. We begin by centering a disk at each vertex, including the corners of R. At vertex v, we place a disk of radius one-half the distance from v to the nearest edge of PR not incident to v. We introduce a subdivision vertex (a degree-2 vertex with a straight angle) at each intersection of a disk boundary and an edge of PR.

Now consider the edges of (the modified) PR that are not covered by disks. Call such an edge *crowded* if its diameter disk intersects the diameter disk of another edge of PR. We mark each crowded edge, and then split each crowded edge by adding its midpoint. We continue marking and splitting in any order until no edges of PR are crowded. We then add the diameter disk of each PR edge so that each edge is a union of diameters of disks as required. Strictly speaking, only the edges of P need be covered by disks, but we include the boundary of R for the sake of neatness.

Next we add disks until all gaps between disks are either 3-gaps or 4-gaps. This can be done by computing the Voronoi diagram of the disks placed so far, and repeatedly placing a maximal-radius disk at a Voronoi vertex and then updating the Voronoi diagram. Bern *et al.* [1] give an $O(n \log^2 n)$ algorithm and Eppstein [9] an $O(n \log n)$ algorithm, where n denotes the number of disks.

Figure 1(a) gives an example of disk-packing, not precisely the same as the one that would be computed by the algorithm just sketched. By adding edges

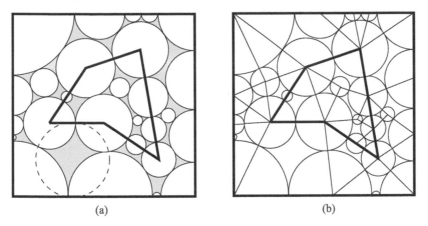

(a)   (b)

Figure 1. (a) A disk-packing respecting the boundary of the polygon. Vertices of 4-gaps are co-circular. (b) Induced triangles and quadrilaterals.

between the centers of tangent disks, the disk-packing induces a decomposition of R into triangles and quadrilaterals as shown in Figure 1(b).

3   Molecules

A *molecule* is a (typically flat) folding of a polygon that can be used as a building block in larger origamis. We shall fold the triangles in the decomposition of R with *rabbit-ear molecules*. In the rabbit-ear molecule, a mountain fold meets each of the triangle's vertices; these folds lie along the angle bisectors of the triangle so that the boundary of the triangle is co-planar in the folded "starfish". A valley fold meets each of the triangle's sides at the points of tangency of the disks; these all fold to a vertical *spine*, perpendicular to the original plane of the paper. The meeting point of the six folds, which becomes the tip of the spine in the folded configuration, is the in-center of the original triangle.

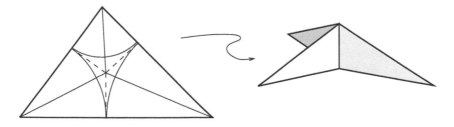

Figure 2. A rabbit-ear molecule folds into a three-armed "starfish".

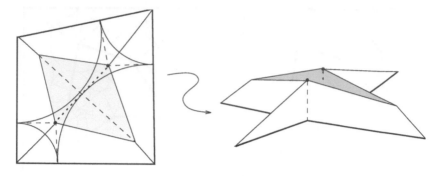

Figure 3. We fold a quadrilateral into a four-armed starfish with a central valley.

At this point, we regard the orientations of the valley folds as changeable: In the larger origami, some of them may be reversed from their initial assignment. For example, to form a flat origami from a single rabbit-ear, one could reverse one of the valleys into a mountain in order to satisfy Maekawa's Theorem.[1] The arms of the starfish all point the same direction away from the spine in the flat origami, and the boundary of the original triangle is collinear.

We shall fold the quadrilaterals as shown in Figure 3. This folding is an improvement, suggested by Robert Lang, of our original method of folding quadrilaterals [14]. In this *gusset molecule* [13], mountain folds extend some distance along the angle bisectors to a *gusset*, a quadrilateral inside the original quadrilateral, shown shaded in Figure 3. The gusset is triangulated with one of its two diagonals, a valley fold, and each of the halves of the overall quadrilateral is folded in a sort of rabbit-ear molecule. This folding of the quadrilaterals enjoys the same property as the folding of the triangles: The valley folds from points of tangency all meet at a central spine, perpendicular to the plane of the paper. Again we regard the orientations of these folds as changeable. In the larger origami, we may reverse one of the valleys in order to form a flat-folding with all arms pointing in the same direction. Notice that such a reversal also sends a crease (a mountain-valley, two-edge path, shown dotted in Figure 3) across the central gusset.

Two of the vertices of the gusset, shown by dots in Figure 3, are fixed by the requirement that valley folds extend perpendicularly from the points of tangency. We refer to these vertices as the *perpendicular points*. The other two vertices of the gusset are not completely constrained. They must, however, lie on the angle bisectors of the quadrilateral in order for the boundary of the quadrilateral to fold to a common plane.

---

[1] Maekawa's Theorem for flat origami [4, 12] states that, at any vertex interior to the paper, the number of mountains minus the number of valleys must be plus or minus two.

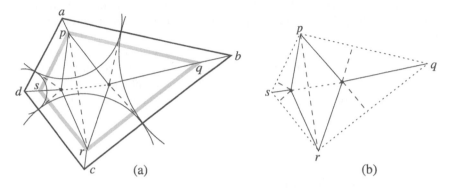

Figure 4. (a) The two unconstrained vertices of the gusset may be chosen to lie on an inset quadrilateral. (b) The inset quadrilateral is folded with two rabbit-ear molecules.

A nice way [13] to locate the the unconstrained vertices—p and r in Figure ??(a)—is to place them at the vertices of an *inset quadrilateral*, a quadrilateral inside the overall quadrilateral, with sides parallel and equidistant to the sides of the original quadrilateral. In Figure ??(a) the original quadrilateral is *abcd* and the inset quadrilateral is *pqrs*. When the gusset molecule is folded, the inset quadrilateral will form a small starfish whose central valley exactly reaches "sea level," that is, *pqrs* and *pr* fold to the same plane. In fact, the gusset folding restricted to *pqrs* is just two rabbit-ear molecules, as shown in Figure ??(b). Hence, the perpendicular points must lie at the in-centers of triangles *pqr* and *prs*, and this requirement determines the size of *pqrs*.

We now argue that all quadrilaterals induced by 4-gaps—all the quadrilaterals that we use—can be folded with the gusset molecule. What we must show is that the triangles *pqr* and *prs* with in-centers at the perpendicular points do, indeed, lie within *abcd*; in other words, that the requirements of the gusset molecule are not in conflict with each other.

First assume that the perpendicular points are distinct, and consider the line L through the perpendicular points. Line L is the line of equal power distance[2] from the disks centered at a and c, and hence passes between these disks. The bisector of the angle between L and the valley fold perpendicular to *bc* fixes r. Since L passes above the disk at c, r lies above c along the angle bisector at c. Thus *pqrs* does, indeed, lie within *abcd*. In the extreme case that the disks at a and c touch each other, *pqrs* equals *abcd* and the gusset molecule reduces to two rabbit-ear molecules.

---

[2]The power distance [1] from a point to a circle is the square of the usual distance minus the radius of the circle squared. For points outside the circle, it is the same as the tangential distance to the circle squared.

What if the perpendicular points coincide? For this extreme case, we use a special property [1] of 4-gaps: The points of tangency of four disks, tangent in a cycle, are co-circular. Figure 1(a) shows the circle for one 4-gap. This property implies that the angle bisectors of the quadrilateral all meet at a common point o, namely the center of the circle through the tangencies. So in the extreme case that the perpendicular points coincide, $pqrs$ shrinks to point o, and the valleys from the points of tangency and the mountains along the angle bisectors all meet at one flat-foldable point.

4   Joining Molecules

We now show how to assign final orientations to creases, so that neighboring molecules fit together and each vertex satisfies Maekawa's Theorem. This fills in (a special case of) a missing step in Lang's Algorithm [13].

We are aiming for a final folding of R that resembles a book of flaps, something like the rightmost picture in Figure 6. More precisely, the folding will look like two books of triangular flaps, one above and one below the original plane of the paper. The molecules (triangles and quadrilaterals) inside P will form the top book, whereas those outside P will form the bottom book. The boundary of P itself will not be folded, and the polygons crossing the boundary, each containing a triangle from two different original molecules, will thus belong to both books.

Angle bisector edges inside P will be mountains and those outside P will be valleys. Other edges of the crease pattern receive *default orientations*, subject to reversal in a final matching step. The default orientation of a *tangency edge* (an edge to a point of tangency) or a *side edge* (an edge along the side of a triangle or quadrilateral) is valley inside P and mountain outside P. Side edges lying along the boundary of P are not folded at all.

At this point, each vertex of the crease pattern has an equal number of mountains and valleys. The vertices interior to R inside P need one more mountain, whereas those outside P need one more valley, in order for molecules to fold to their assigned half-spaces, above or below the original plane of the paper. (Vertices on the boundary of P can have an excess of either mountains or valleys.) Let G be the planar graph obtained from the decomposition by removing all angle bisector edges and all edges along the boundary of P. We would like to find a set of edges M—a matching—such that each vertex of G lying in the interior of R is incident to exactly one edge of M. By reversing the orientations of the edges of M, we ensure that each vertex satisfies Maekawa's

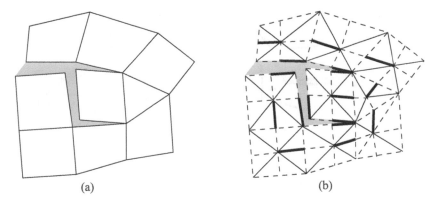

(a)   (b)

Figure 5. (a) Cutting out a tree T_C (shaded) spanning interior corners leaves a tree of molecules T_M. Roots are at the upper left. (b) The matching consists of side edges from corners to parents in T_C and tangency edges from molecule centers to parents in T_M. Assignments shown assume all molecules are inside P.

Theorem. All vertices, even the ones along P, which lost two edges each from the original decomposition of R, also satisfy Kawasaki's Theorem.[3]

We now show how to solve the matching problem using dual spanning trees. Let T_C be a tree of side edges such that: T_C includes no edges along the boundary of R or P; T_C spans all interior corners of molecules; and T_C spans exactly one corner along the boundary of R, which we consider to be its root. If we were to cut the paper along T_C, we would obtain a tree of molecules T_M, as shown in Figure 5(a). We root T_M at one of the molecules incident to the root of T_C. The matching M contains two types of edges: each tangency edge from the center of a molecule to the side of its parent in T_M (along with one such edge inside the root molecule), and each side edge from a corner to its parent (a tangency point) in tree T_C. See Figure 5(b).

To picture the effect of this choice of M on the eventual flat-folding, imagine that we have actually cut along the edges of T_C. Imagine building up the flat-folding molecule-by-molecule in a pre-ordered traversal of T_M. The root molecule of T_M folds to a book of flaps with collinear edges lying along the original plane of the paper. Each child molecule adds a "pamphlet" of three or four flaps between two flaps of the book we have constructed so far. The cover and back cover of the pamphlet are glued to their adjacent pages, so that a quadrilateral thickens two old flaps and adds two new flaps.

We continue gluing pamphlets between flaps of the growing book as we

---

[3]Kawasaki's Theorem for flat origami [4, 12] states that at any vertex interior to the paper, the sum of alternate angles must be 180°.

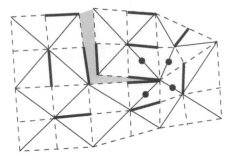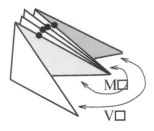

Figure 6. Taping together a cut leading to a leaf of T_C amounts to joining two "armpits" in the book of flaps.

go down the tree. Whenever we cross the boundary of P, we glue the next pamphlet above or below—rather than between flaps of—its parent molecule, so that the boundary of P is not folded itself. When we are done joining all the molecules, we have two books of flaps, one above and one below the original plane of the paper.

Now imagine taping the cut edges back together in a postorder traversal of T_C. Before taping, the cut leading to a leaf of T_C, say inside P, defines the bottom edge of two adjacent "armpits", as shown in Figure 6. (An armpit consists of one layer from each of two adjacent flaps.) Taping together the first and last layer of the intervening flap forms a mountain fold, agreeing with the orientation we gave to side edges in the matching. Taping together the remaining two sides of the cut forms a valley fold, agreeing with the default orientation of side edges inside P. Taping a cut leading to a leaf of T_C closes two armpits and reduces the number of flaps in the book by two. We can continue taping cuts all the way up T_C. Since each taping joins armpits adjacent at the time of the taping, there can be no "crossed" pair of tapings, or put another way, no place where the paper is forced to penetrate itself. Altogether the taping completes a crease pattern on paper R that can be folded flat so that P lies above, and its complement $R \setminus P$ lies below, the original plane of the paper.

5   Fattening the Polygon

At this point, we have a degenerate solution to the cut-out problem. A cut through the original plane of the paper separates P from its complement. Unfortunately, it also cuts P into its constituent molecules. A cut very slightly below the original plane of the paper leaves P intact, while adding a small

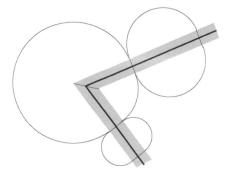

Figure 7. Fattening the polygon into a ribbon lets P survive the cut intact.

"rim" to P.

We can remove the degeneracy by fattening the boundary of P into a narrow "ribbon" as shown in Figure 7. The boundaries of the ribbon are slightly inside and outside the original P; vertices of P are moved in or out along angle bisectors. (We actually saw this ribbon construction already: A gusset molecule is two adjacent rabbit-ear molecules surrounded by a ribbon!) The width of the ribbon must be smaller than the *minimum feature size* of the polygon, the minimum distance between a vertex of P and an edge not incident to that vertex.

We modify the disk-packing step so that it packs partial disks (sectors) around the boundary of ribbon, such that interior and exterior sectors match up. Creases between corresponding subdivision points cross the ribbon at right angles, whereas creases between corresponding vertices cross at angle bisectors, so that each vertex still satisfies Kawasaki's Theorem. Notice that interior and exterior sectors centered on corresponding vertices have slightly different radii.

6   Discussion

We have given an algorithm for the cut-out problem. More precisely, we have given an algorithm for computing a crease pattern with a flat-folding that solves the cut-out problem. We have not described how to actually transform the crease pattern into the flat-folding.

Is our algorithm usable? The answer is a qualified yes. Figure 8 gives a crease pattern for a fish cut-out that is not too hard to fold. In this crease pattern, we have taken a number of shortcuts to make the algorithm more practical. First, we have used only three-sided and special four-sided gaps, ones in which perpendiculars from the center vertex o happen to meet the sides

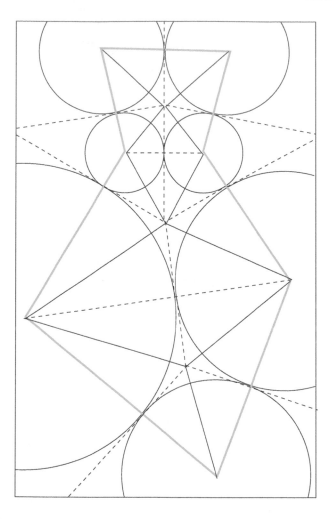

Figure 8. An example for the reader to try. This "mounted marlin" design incorporates some practical shortcuts. For example, packing the exterior with disks is unnecessary, because radiating folds do not collide.

at the points of tangency. Second, we have not packed the disks all the way to the boundary of the paper, only far enough that radiating folds do not meet within the page. Third, we have not fattened the polygon, and hence, the cut should be placed slightly below the original plane of the paper, so that the interior remains connected.

The number of creases used by our algorithm is not really excessive, linear in the number of disks in the initial disk-packing. The number of disks, in turn, depends upon a fairly natural complexity measure of the polygon. Define the *Local Feature Size*, $LFS(p)$, at a point p on an edge e of P to be the distance to the closest edge that is not adjacent to e [16]. The local feature size is small at narrow necks of the polygon. It is not hard to see that the number of disks around the boundary of P is $O(\int_{\partial P} 1/|LFS(p)|)$, where the integral is over the boundary of P. The number of additional disks needed to fill out the square is linear in the disks around the boundary of P, because each new disk reduces the number of sides of the gap into which it is placed.

The algorithm of this paper can be generalized to the problem in which the input is a planar straight-line graph G, and a single cut must cut along all the edges of G. An interesting open question asks whether there is a polynomial-size solution (polynomial in the number of original vertices of P or G) for the cut-out problem. A solution using disk-packing may shed some light on two other computational geometry problems: simultaneous inside-outside nonobtuse triangulation [2] and conforming Delaunay triangulation [8].

Acknowledgments

We would like to thank Robert Lang for suggesting the gusset molecule; this suggestion significantly improved our results.

References

[1] M. Bern, S. Mitchell, and J. Ruppert. Linear-size nonobtuse triangulation of polygons. *Disc. Comput. Geom.* 14 (1995) 411–428.

[2] M. Bern and D. Eppstein. Polynomial-size nonobtuse triangulation of polygons. *Int. J. of Comp. Geom. and Applications* 2 (1992) 241–255.

[3] M. Bern and D. Eppstein. Quadrilateral meshing by circle packing. *Proc. 6th Int. Meshing Roundtable*, 1997, 7–19.

[4] M. Bern and B. Hayes. The complexity of flat origami. *Proc. 7th ACM-SIAM Symp. Disc. Algorithms*, 1996, 175–183.

[5] E. Demaine and M. Demaine. Computing extreme origami bases. Tech. Report CS-97-22, Dept. of Computer Science, U. of Waterloo, May 1997.

[6] E. Demaine, M. Demaine, and A. Lubiw. Folding and cutting paper. *Revised papers from the Japan Conference on Discrete and Computational Geometry*, J. Akiyama, M. Kano, and M. Urabe, eds., Tokyo, 1998. Lecture Notes in Computer Science, Vol. 1763, Springer, 104–117.

[7] E. Demaine, M. Demaine, and A. Lubiw. Folding and one straight cut suffice. *Proc. 10th ACM-SIAM Symp. Disc. Algorithms*, 1999, 891–892.

[8] H. Edelsbrunner and T.-S. Tan. An upper bound for conforming Delaunay triangulations. *Disc. Comput. Geom.* 10 (1993) 197–213.

[9] D. Eppstein. Faster circle packing with application to nonobtuse triangulation. *Int. J. Comp. Geom. & Applications* 7 (1997) 485–491.

[10] M. Gardner. Paper Cutting. In *New Mathematical Diversions*, Math. Assoc. America, 1995, 58–69.

[11] H. Houdini. In *Paper Magic*, E. P. Dutton & Company, 1922, 176–177.

[12] T. Hull. On the mathematics of flat origamis. *Congressus Numerantium* 100 (1994) 215–224.

[13] R. Lang. A computational algorithm for origami design. *Proc. 12th ACM Symp. Comp. Geometry*, 1996, 98–105.

[14] E. Lodi, L. Pagli, and N. Santoro, eds. *Fun with Algorithms*, Carleton Scientific, 1999.

[15] National Standards and Emblems. *Harper's New Monthly Magazine* 47 (1873) 171–181.

[16] J. Ruppert. A new and simple algorithm for quality 2-dimensional mesh generation. *Proc. 4th ACM-SIAM Symp. Disc. Algorithms*, 1993, 83–92.

The Combinatorics of
Flat Folds: A Survey

Thomas C. Hull

1   Introduction

It is safe to say that in the study of the mathematics of origami, flat origami has received the most attention. To put it simply, a *flat origami model* is one which can be pressed in a book without (in theory) introducing new creases. We say "in theory" because when one actually folds a flat origami model, slight errors in folding will often make the model slightly nonflat. In our analysis, however, we ignore such errors and assume all of our models are perfectly folded. We also assume that our paper has zero thickness and that our creases have no width. It is surprising how rich the results are using a purely combinatorial analysis of flat origami. In this paper, we introduce the basics of this approach, survey the known results, and briefly describe where future work might lie.

First, some basic definitions are in order. A *fold* refers to any folded paper object, independent of the number of folds done in sequence. The *crease pattern* of a fold is a planar embedding of a graph which represents the creases that are used in the final folded object. (This can be thought of as a structural blueprint of the fold.) Creases come in two types: *mountain creases*, which are convex, and *valley creases*, which are concave (see Figure 1). Clearly the type of a crease depends on which side of the paper we look at, and so we assume we are always looking at the same side of the paper.

We also define a *mountain-valley (MV) assignment* to be a function mapping the set of all creases to the set $\{M, V\}$. In other words, we label each crease

Figure 1. Mountain creases, valley creases, and the crease pattern for the flapping bird with MV assignment shown.

mountain or valley. MV assignments that can actually be folded are called *valid*, while those which do not admit a flat-folding (i.e., force the paper to self-intersect in some way) are called *invalid*.

There are two basic questions on which flat-folding research has focused:

1. Given a crease pattern, without an MV assignment, can we tell whether it can flat-fold?

2. If an MV assignment is given as well, can we tell whether it is valid?

These are also the focus of this survey. We will not discuss the special cases of flat origami tessellations, origami model design, or other science applications.

2   Classic Single Vertex Results

We start with the simplest case for flat origami folds. We define a *single vertex fold* to be a crease pattern (no MV assignment) with only one vertex in the interior of the paper and all crease lines incident to it. Intersections of creases on the boundary of the paper clearly follow different rules, and nothing of interest has been found to say about them thus far (except in origami design; see [14], [15]). A single vertex fold which is known to fold flat is called a *flat vertex fold*. We present a few basic theorems relating to necessary and sufficient conditions for flat-foldability of single vertex folds. These theorems appear in their cited references without proof. While Kawasaki, Maekawa, and Justin undoubtedly had proofs of their own, the proofs presented below appear in [3].

Theorem 2.1 (Kawasaki [10], Justin [5], [6]). *Let v be a vertex of degree 2n in a single vertex fold and let $\alpha_1, ..., \alpha_{2n}$ be the consecutive angles between the creases. Then v is a flat vertex fold if and only if*

$$\alpha_1 - \alpha_2 + \alpha_3 - \cdots - \alpha_{2n} = 0. \tag{1}$$

Proof. Consider a simple closed curve which winds around the vertex. This curve mimics the path of an ant walking around the vertex on the surface of the paper after it is folded. We measure the angles the ant crosses as positive when traveling to the left and negative when walking to the right. Arriving at the point where the ant started means that this alternating sum is zero. The converse is left to the reader; see [3] for more details. □

Theorem 2.2 (Maekawa, Justin [6], [8]). *Let M be the number of mountain creases and V be the number of valley creases adjacent to a vertex in a single vertex fold. Then $M - V = \pm 2$.*

Proof. (Siwanowicz) If n is the number of creases, then $n = M + V$. Fold the paper flat and consider the cross-section obtained by clipping the area near the vertex from the paper; the cross-section forms a flat polygon. If we view each interior $0°$ angle as a valley crease and each interior $360°$ angle as a mountain crease, then $0V + 360M = (n - 2)180 = (M + V - 2)180$, which gives $M - V = -2$. On the other hand, if we view each $0°$ angle as a *mountain* crease and each $360°$ angle as a *valley* crease (this corresponds to flipping the paper over), then we get $M - V = 2$. □

In the literature, Theorem 2.1 and 2.2 are referred to as Kawasaki's Theorem and Maekawa's Theorem, respectively. Justin [7] refers to equation (1) as the *isometries condition*. Kawasaki's Theorem is sometimes stated in the equivalent form that the sum of alternate angles around v equals $180°$, but this is only true if the vertex is on a flat sheet of paper. Indeed, notice that the proofs of Kawasaki's and Maekawa's Theorems do not use the fact that $\sum \alpha_i = 360°$. Thus these two theorems are also valid for single vertex folds where v is at the apex of a cone-shaped piece of paper. We will require this generalization in Sections 4 and 5.

Note that while Kawasaki's Theorem assumes that the vertex has even degree, Maekawa's Theorem does not. Indeed, Maekawa's Theorem can be used to prove this fact. Let v be a single vertex fold that folds flat and let n be the degree of v. Then $n = M + V = M - V + 2V = \pm 2 + 2V$, which is even.

3 Generalizing Kawasaki's Theorem

Kawasaki's Theorem gives us a complete description of when a single vertex in a crease pattern will (locally) fold flat. Figure 2 shows two examples of crease patterns which satisfy Kawasaki's Theorem at each vertex, but which will not fold flat. The example on the left is from [3], and a simple argument

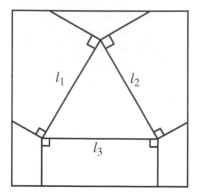 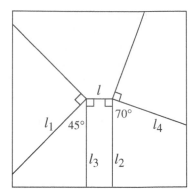

Figure 2. Two impossible-to-fold-flat folds.

shows that no two of the creases l_1, l_2, l_3 can have the same MV parity. Thus no valid MV assignment for the lines l_1, l_2, l_3 is possible. The example on the right has valid MV assignments, but still fails to fold flat. The reader is encouraged to copy this crease pattern and try to fold it flat, which will reveal that some flap of paper will have to intersect one of the creases. However, if the location of the two vertices is changed relative to the border of the paper, or if the crease l is made longer, then the crease pattern *will* fold flat.

This illustrates how difficult the question of flat-foldability is for multiple vertex folds. Indeed, in 1996, Bern and Hayes [1] proved that the general question of whether or not a given crease pattern can fold flat is NP-hard. Thus one would not expect to find easy necessary and sufficient conditions for general flat-foldability.

We will present two efforts to describe general flat-foldability. The first has to do with the realization that when we fold flat along a crease, one part of the paper is being reflected along the crease line to the other side. Let us denote $R(l_i)$ to be the reflection in the plane, $\mathbb{R}^2$, along a line l_i.

Theorem 3.1 (Kawasaki [9], [12], Justin [5], [7]). *Given a multiple vertex fold, let γ be any closed, vertex-avoiding curve drawn on the crease pattern which crosses crease lines $l_1, ..., l_n$, in order. Then, if the crease pattern can fold flat, we will have*

$$R(l_1)R(l_2)\cdots R(l_n) = I \tag{2}$$

where I denotes the identity transformation.

Although a rigorous proof of Theorem 3.1 does not appear in the literature, we sketch here a proof by induction on the number of vertices. In the base case, we are given a single vertex fold, and it is a fun exercise to show that condition (2) is equivalent to Equation (1) in Kawasaki's Theorem (use the fact that the composition of two reflections is a rotation). The induction step then proceeds by breaking the curve γ containing k vertices into two closed curves, one containing $k - 1$ vertices and one containing a single vertex (the kth).

The condition (2) is not a sufficient condition for flat-foldability (the crease patterns in Figure 2 are counterexamples here as well). In fact, as the induction proof illustrates, Theorem 3.1 extends Kawasaki's Theorem to as general a result as possible.

In [7], Justin proposes a necessary and sufficient condition for general flat-foldability, although as Bern and Hayes predicted, it is not very computationally feasible. To summarize, let C be a crease pattern for a flat origami model, but for the moment we are considering the boundary of the paper as part of the graph. If E denotes the set of edges in C embedded in the plane, then we call $\mu(E)$ the *f-net*, which is the image of all creases and boundary of the paper after the model has been folded. We then call $\mu^{-1}(\mu(E))$ the *s-net*. This is equivalent to imagining that we fold carbon-sensitive paper, rub all the crease lines firmly, and then unfold. The result will be the *s*-net.

Justin's idea is as follows: Take all the faces of the *s*-net which get mapped by μ to the same region of the *f*-net and assign a *superposition order* to them in accordance to their layering in the final folded model. One can thus try to fold a given crease pattern by cutting the piece of paper along the creases of the *s*-net, tranforming them under μ, applying the superposition order, and then attempting to glue the paper back together. Justin describes a set of three intuitive *crossing conditions* (see [7]) which must not happen along the *s*-net creases during the glueing process if the model is to be flat-foldable—if this can be done, we say that the *noncrossing condition* is satisfied. Essentially Justin conjectures that a crease pattern folds flat if and only if the noncrossing condition holds. Although the spirit of this approach seems to accurately reflect the flat-foldability of multiple vertex folds, no rigorous proof appears in the literature; it seems that this is an open problem.

4   Generalizing Maekawa's Theorem

To extend Maekawa's Theorem to more than one vertex, we define interior vertices in a flat multiple vertex fold to be *up vertices* and *down vertices* if they locally have $M - V = 2$ or -2, respectively. We define a crease line to be an *interior* crease if its endpoints lie in the interior of the paper (as opposed to on the boundary), and consider any crease line with both endpoints on the boundary of the paper to actually be two crease lines with an interior vertex of degree 2 separating them.

Theorem 4.1 (Hull [3]). *Given a multiple vertex flat-fold, let M (resp. V) denote the number of mountain (resp. valley) creases, U (resp. D) denote the number of up (resp. down) vertices, and M_i (resp. V_i) denote the number of interior mountain (resp. valley) creases. Then*

$$M - V = 2U - 2D - M_i + V_i.$$

Another interesting way to generalize Maekawa's Theorem is to explore restrictions that turn it into a sufficiency condition. In the case where all of the angles around a single vertex are equal, an MV assignment with $M - V = \pm 2$ is guaranteed to be valid. This observation can be generalized to sequences of consecutive equal angles around a vertex.

Let us denote a single vertex fold by $v = (\alpha_1, ..., \alpha_{2n})$ where the α_i are consecutive angles between the crease lines. We let $l_1, ..., l_{2n}$ denote the creases adjacent to v where α_i is the angle between creases l_i and l_{i+1} (α_{2n} is between l_{2n} and l_1).

If $l_i, ..., l_{i+k}$ are consecutive crease lines in a single vertex fold which have been given a MV assignment, let $M_{i,...,i+k}$ be the number of mountains and $V_{i,...,i+k}$ be the number of valleys among these crease lines. We say that a given MV assignment is valid for the crease lines $l_i, ..., l_{i+k}$ if the MV assignment can be followed to fold up these crease lines without forcing the paper to self-intersect. (Unless these lines include all the creases at the vertex, the result will be a cone.) The necessity portion of the following result appears in [4], while sufficiency is new.

Theorem 4.2. *Let $v = (\alpha_1, ..., \alpha_{2n})$ be a single vertex fold in either a piece of paper or a cone, and suppose we have $\alpha_i = \alpha_{i+1} = \alpha_{i+2} = \cdots = \alpha_{i+k}$ for some $i < k \leq 2n$. Then a given MV assignment is valid for $l_i, ..., l_{i+k+1}$ if and only if*

$$M_{i,...,i+k+1} - V_{i,...,i+k+1} = \begin{cases} 0 & \text{when } k \text{ is even} \\ \pm 1 & \text{when } k \text{ is odd.} \end{cases}$$

Proof. Necessity follows by applications of Maekawa's Theorem. If k is even, then the cross-section of the paper around the creases in question might appear as shown in the left of Figure 3. If we consider only this sequence of angles and imagine adding a section of paper with angle β to connect the loose ends at the left and right (see Figure 3, left), then we'll have a flat-folded cone which must satisfy Maekawa's Theorem. The angle β adds two extra creases, both of which must be mountains (or valleys). We may assume that the vertex points

Figure 3. Applying Maekawa when k is even (left) and odd (right).

up, and thus we subtract two from the result of Maekawa's Theorem to get $M_{i,...,i+k+1} - V_{i,...,i+k+1} = 0$.

If k is odd (Figure 3, right), then this angle sequence, if considered by itself, will have the loose ends from angles α_{i-1} and α_{i+k+1} pointing in the same direction. If we glue these together (extending them if necessary), then Maekawa's Theorem may be applied. After subtracting (or adding) one to the result of Maekawa's Theorem, because of the extra crease made when gluing the loose flaps, we get $M_{i,...,i+k+1} - V_{i,...,i+k+1} = \pm 1$.

For sufficiency, we proceed by induction on k. The result is trivial for the base cases $k = 0$ (only one angle, and the two neighboring creases will either be M, V or V, M) and $k = 1$ (two angles, and all three possible ways to assign 2 Ms and 1 V, or vice-versa, can be readily checked to be foldable). For arbitrary k, we will always be able to find two adjacent creases l_{i+j} and l_{i+j+1} to which the MV assignment assigns opposite parity. Let l_{i+j} be M and l_{i+j+1} be V. We make these folds and can imagine that α_{i+j-1} and α_{i+j} have been fused into the other layers of paper, i.e., removed. The value of $M - V$ will not have changed for the remaining sequence $l_i, ..., l_{i+j-1}, l_{i+j+2}, ..., l_{i+k}$ of creases, which are flat-foldable by the induction hypothesis. $\square$

5   Counting Valid MV Assignments

We now turn to the question of counting the different ways we can fold a flat origami model. By this we mean, given a crease pattern that is known to fold flat, how many different valid MV assignments are possible?

We start with the single vertex case. Let $C(\alpha_1, ..., \alpha_{2n})$ denote the number of valid MV assignments possible for the vertex fold $v = (\alpha_1, ..., \alpha_{2n})$.

An an example, consider the case where $n = 2$ (so we have four crease lines at v). We compute $C(\alpha_1, \alpha_2, \alpha_3, \alpha_4)$ using Maekawa's Theorem. Its value will depend on the type of symmetry the vertex has, and the three possible situations are depicted in Figure 4. $C(90, 90, 90, 90) = 8$ because any crease could be the "odd crease out" and the vertex could be up or down. In Figure 4(b) we have only mirror symmetry, and by Theorem 4.2, $M_{2,3,4} - V_{2,3,4} = \pm 1$. Thus l_2, l_3, l_4 must have two Ms and one V or vice versa; this determines l_1's parity, giving $C(\alpha_1, ..., \alpha_4) = 6$. In Figure 4(c) $M_{1,2} - V_{1,2} = 0$, so l_1 and

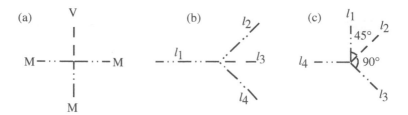

Figure 4. The three scenarios for vertices of degree 4.

l_2 can be M,V or V,M, and the other two must be both M or both V, giving $C(\alpha_1, ..., \alpha_4) = 4$.

The example in Figure 4(a) represents the case with no restrictions. This appears whenever all the angles are equal around v, giving

$$C(\alpha_1, ..., \alpha_{2n}) = 2 \binom{2n}{n-1}.$$

The idea in Figure 4(c), where we pick the smallest angle we see and let its creases be M,V or V,M, can be applied inductively to give the lower bound in the following (see [4] for a full proof).

Theorem 5.1. *Let* $v = (\alpha_1, ..., \alpha_{2n})$ *be the vertex in a flat vertex fold, on either a flat piece of paper or a cone. Then*

$$2^n \le C(\alpha_1, ..., \alpha_{2n}) \le 2 \binom{2n}{n-1}$$

are sharp bounds.

A formula for $C(\alpha_1, ..., \alpha_{2n})$ seems out of reach, but using the equal-angles-in-a-row concept, recursive formulas exist to compute this quantity in linear time.

Theorem 5.2 (Hull [4]). *Let* $v = (\alpha_1, ..., \alpha_{2n})$ *be a flat vertex fold in either a piece of paper or a cone, and suppose we have* $\alpha_i = \alpha_{i+1} = \alpha_{i+2} = \cdots = \alpha_{i+k}$ *and* $\alpha_{i-1} > \alpha_i$ *and* $\alpha_{i+k+1} > \alpha_{i+k}$ *for some* i *and* k. *Then*

$$C(\alpha_1, ..., \alpha_{2n}) = \binom{k+2}{\frac{k+2}{2}} C(\alpha_1, ..., \alpha_{i-2}, \alpha_{i-1} - \alpha_i + \alpha_{i+k+1}, \alpha_{i+k+2}, ..., \alpha_{2n})$$

if k *is even, and*

$$C(\alpha_1, ..., \alpha_{2n}) = \binom{k+2}{\frac{k+1}{2}} C(\alpha_1, ..., \alpha_{i-1}, \alpha_{i+k+1}, ..., \alpha_{2n})$$

if k *is odd.*

Theorem 5.2 was first stated in [4], but the basic ideas behind it are discussed by Justin in [7].

As an example, consider $C(20, 10, 40, 50, 60, 60, 60, 60)$. Theorem 5.1 tells us that this quality lies between 16 and 112. But using Theorem 5.2, we see that $C(20, 10, 40, 50, 60, 60, 60, 60) = \begin{pmatrix} 2 \\ 1 \end{pmatrix} C(50, 50, 60, 60, 60, 60) = \begin{pmatrix} 2 \\ 1 \end{pmatrix} \begin{pmatrix} 3 \\ 1 \end{pmatrix} C(60, 60, 60, 60) = \begin{pmatrix} 2 \\ 1 \end{pmatrix} \begin{pmatrix} 3 \\ 1 \end{pmatrix} 2 \begin{pmatrix} 4 \\ 1 \end{pmatrix} = 48$.

Not much is known about counting valid MV assignments for flat multiple vertex folds. While there are similarities to work done on counting the number of ways to fold up a grid of postage stamps (see [13], [16], [17]), the questions asked are slightly different. For other work, see [7] and [4].

6 Conclusion

The results for flat-foldability seem to have almost exhausted the single vertex case. Open problems exist, however, in terms of global flat-foldability, and very little is known about enumerating valid MV assignments for multiple vertex crease patterns.

References

[1] Bern, M. and Hayes, B., "The complexity of flat origami", *Proceedings of the 7th Annual ACM-SIAM Symposium on Discrete Algorithms*, (1996) 175–183.

[2] Ewins, B. and Hull, T., personal communication, 1994.

[3] Hull, T., "On the mathematics of flat origamis", *Congressus Numerantium*, 100 (1994) 215-224.

[4] Hull, T., "Counting mountain-valley assignments for flat folds", *Ars Combinatoria*, to appear.

[5] Justin, J., "Aspects mathematiques du pliage de papier" (in French), in: H. Huzita ed., *Proceedings of the First International Meeting of Origami Science and Technology*, Ferrara, (1989) 263-277.

[6] Justin, J., "Mathematics of origami, part 9", *British Origami* (June 1986) 28-30.

[7] Justin, J., "Toward a mathematical theory of origami", in: K. Miura ed., *Origami Science and Art: Proceedings of the Second International Meeting of Origami Science and Scientific Origami*, Seian University of Art and Design, Otsu, (1997) 15-29.

[8] Kasahara, K. and Takahama, T., *Origami for the Connoisseur*, Japan Publications, New York, (1987).

[9] Kawasaki, T. and Yoshida, M., "Crystallographic flat origamis", *Memoirs of the Faculty of Science, Kyushu University, Series A*, Vol. 42, No. 2 (1988), 153-157.

[10] Kawasaki, T., "On the relation between mountain-creases and valley-creases of a flat origami" (abridged English translation), in: H. Huzita ed., *Proceedings of the First International Meeting of Origami Science and Technology*, Ferrara, (1989) 229-237.

[11] Kawasaki, T., "On the relation between mountain-creases and valley-creases of a flat origami" (unabridged, in Japanese), Sasebo College of Technology Reports, 27, (1990) 55-80.

[12] Kawasaki, T., "$R(\gamma) = I$", in: K. Miura ed., *Origami Science and Art: Proceedings of the Second International Meeting of Origami Science and Scientific Origami*, Seian University of Art and Design, Otsu, (1997) 31-40.

[13] Koehler, J., "Folding a strip of stamps", *Journal of Combinatorial Theory*, 5 (1968) 135-152.

[14] Lang, R.J., "A computational algorithm for origami design", *Proceedings of the 12th Annual ACM Symposium on Computational Geometry*, (1996), 98-105.

[15] Lang, R.J., *TreeMaker 4.0: A Program for Origami Design*, (1998) http://origami.kvi.nl/programs/TreeMaker/trmkr40.pdf

[16] Lunnon, W.F., "A map-folding problem", *Mathematics of Computation*, 22, No. 101 (1968) 193-199.

[17] Lunnon, W.F., "Multi-dimensional map folding", *The Computer Journal*, 14, No. 1 (1971) 75-80.

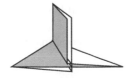

A Mathematical Model for Non-Flat Origami

sarah-marie belcastro and Thomas C. Hull

This project was inspired by the intuition that origami must have some significant mathematical properties. Much of the origami commonly done in the United States folds flat; some origami is then made three-dimensional by manipulating the flat-folded paper to look more lifelike. The most common examples of this phenomenon are the flapping bird and the water bomb. Flat-foldability has been analyzed in [2] and [4]. However, there is nothing either artistically or mathematically which restricts one to folding paper flat. There are origami designs (though comparatively few) which require non-flat folds—creases where the dihedral angle is neither 0 nor 2π—for example, the traditional Masu box.

In this paper, we will assume that any fold is not necessarily flat. We will model this type of folding and show necessary conditions for a crease pattern to be foldable. Sufficiency conditions are significantly more difficult and are briefly discussed in [1].

1 Modeling Paper-Folding: Constraints

We wish to model the folding of paper mathematically, by examining a map $f : \mathbb{R}^2 \to \mathbb{R}^3$. In fact, we will use a composition map

$$\mathbb{R}^2 \hookrightarrow \mathbb{R}^3 \to \mathbb{R}^3;$$

$$(x,y) \mapsto (x,y,0) \mapsto (?,?,?).$$

Because we think of the paper as sitting in $\mathbb{R}^3$ in this fashion, we will henceforth examine only the $\mathbb{R}^3 \to \mathbb{R}^3$ portion of the above map.

Consider a folded piece of paper; mentally unfold it. As a combinatorial object, it has vertices, edges, and faces. While curved creases are, of course, possible (see [3]), in this paper we will only consider straight creases.

Definition: A *crease pattern* is a simple straight-edged graph embedded in the plane. The edges of this graph correspond to the locations of fold lines in an unfolded sheet of paper. Note that we do not include the boundary of the paper in the crease pattern.

In the most rudimentary way, we may begin by thinking of the mechanics of putting a crease into paper. In reality, we hold part of the paper stationary, and rotate the rest of the paper by π radians around the proposed crease line. We then crease the paper; if we desire the result to be nonflat, then we change the dihedral angle δ of the crease to be greater than zero.

Mathematically, we can accomplish the same result by applying a rotation by $\pi - \delta$ radians, which we call the *folding angle*, to the portion of the plane corresponding to the part of the paper we lifted to make our crease. This sets the stage for the following analysis.

There are several properties that folded paper has which we will need to encode:

(1) Paper does not stretch.

(2) The faces of the folded paper are flat (as opposed to curved).

(3) We do not want the paper to rip or have holes.

(4) Paper does not intersect itself.

Constraint (2) indicates that f must be piecewise affine; constraint (1) restricts f to a piecewise isometry. In fact, as we observed above, f will be composed only of rotation matrices. Constraint (3) specifies that f must be continuous. Constraint (4) is at the heart of any sufficiency condition, and is as yet not completely accounted for—see [1, Section 1.5].

Definition: A crease pattern is *foldable* if it is physically possible to produce a piece of paper which has crease lines in the indicated positions with dihedral angles as indicated, such that the paper neither stretches, rips, nor self-intersects during folding and such that the faces of the resulting fold are flat.

For simplicity, we first consider crease patterns with only a single multi-valent vertex. (We will abuse notation by using the term *single-vertex fold*.)

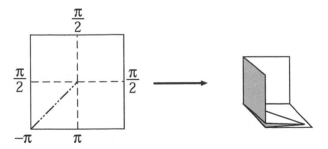

Figure 1. A labeled crease pattern and the corresponding completed fold. Note that only the folding angles for the creases are labeled.

The reader may wish to think of this as a local case, where we examine an open set around a particular vertex of a more complex crease pattern.

2   Analysis of a Single-Vertex Fold

We begin with an example of a single-vertex fold in Figure 1, namely a box-pleated corner. In general, the crease lines $l_1, ..., l_n$ are enumerated in counter-clockwise order beginning from the x-axis. We label each crease line l_i with a pair of angles (α_i, ρ_i) where α_i denotes the position of the crease in the paper (the plane angle) and ρ_i denotes the folding angle. However, in our illustrations, we only label the folding angles because the plane angles are implicit from the drawing of the crease pattern.

We wish to define our map f using only the given information $\{l_i : (\alpha_i, \rho_i)\}$. To ensure continuity of the map, we impose a folding order via a closed path around the vertex.

A spider[1] crawls around the vertex counterclockwise, beginning just below the x-axis.[2] If the spider[3] is sufficiently two-dimensional, it can also traverse this path when the paper is folded. We wish to describe the journey of the spider in a mathematically formal way.

To travel the crease pattern (resp. on a completed fold), the spider takes two actions:

---

[1] ...or ant, or planarian, or triceratops.

[2] We ignore the possibility that the spider may lift itself from the paper or choose to spin a web rather than crawling.

[3] ...or ant, or planarian, or triceratops... We note that after this paper was presented at the conference, Robert Lang produced a large teal planarian for our use.

(1) It moves across a face of the crease pattern (resp. across a face of the completed fold). This will be accounted for by a change in the xy-coordinates on the plane of the unfolded paper.

(2) It moves from from one face to another by crossing a crease line l_i (resp. by rotating around a crease line l_i). In the completed fold, the spider rotates by the supplement ρ_i of the dihedral angle δ_i between the two faces. Because the fold is a piecewise linear object, we may express this action via a linear map (matrix), and denote it by L_i.

Consider, then, a single-vertex crease pattern on which each crease line l_i is assigned an ordered pair (α_i, ρ_i).

Let A_i be the matrix corresponding to a rotation in the xy-plane by the plane (or location) angle α_i. Let C_i be the matrix which rotates by the folding angle ρ_i in the yz-plane, i.e., around the x-axis counterclockwise. Now, use $\chi_i = A_i C_i A_i^{-1}$ to denote the rotation counterclockwise of angle ρ_i around the axis corresponding to the crease line l_i in the plane.

Note: A_i and C_i (and thus χ_i as well) are orthogonal matrices, which reflect the nonrubbery nature of the paper.

Now we will define the L_i in terms of the χ_j. Certainly $L_1 = \chi_1$, and $L_2 = \chi_1 \chi_2 \chi_1^{-1}$ because we undo operation L_1 to put crease line l_2 back in the xy-plane, then perform fold χ_2, and redo our first crease to put the whole ensemble back. (Recall that the order of matrix multiplication is determined by viewing the process as linear functions applied to a vector in $\mathbb{R}^3$.) Using this type of reasoning, we see that in general $L_i =$(matrix to redo the previous L's).χ_i.(matrix to undo the previous L's in reverse order), so that as a matrix L_i is equal to $(L_{i-1}...L_1)\chi_i(L_1^{-1}...L_{i-1}^{-1})$. This recursive definition gives us $L_i = \chi_1...\chi_{i-1}\chi_i\chi_{i-1}^{-1}...\chi_1^{-1}$.

Notice that we have implicitly defined a piecewise linear map $f : \mathbb{R}^2 \subset \mathbb{R}^3 \to \mathbb{R}^3$ corresponding to our fold! Enumerate the faces of the crease pattern in $\mathbb{R}^2$ as $F_1, ..., F_n$ so that face F_j lies between crease lines l_j and l_{j+1}. Now, set the spider down on the x-axis and hope it crawls counterclockwise. Explicitly, $f(x, y)$ is

$$f(x, y, 0) = \begin{cases} L_1(x, y, 0) & \text{for } (x, y) \in F_1 \\ L_2 L_1(x, y, 0) & \text{for } (x, y) \in F_2 \\ \vdots & \vdots \\ L_{n-1} \cdots L_1(x, y, 0) & \text{for } (x, y) \in F_{n-1} \\ L_n L_{n-1} \cdots L_1(x, y, 0) = I(x, y, 0) & \text{for } (x, y) \in F_n \end{cases}.$$

Note that the "last" face of the paper lies in the xy-plane, and that this includes both l_1 and l_n. Also, it is easy to redefine the map for a clockwise spider path; we leave this to the reader.

3 Necessity, not Sufficiency

Given a single-vertex crease pattern with n creases, we may write the accumulation of the operations on the spider (after it has traversed its path) as $L_n L_{n-1}...L_2 L_1$.

In fact, we see that we are required to have $L_n L_{n-1}...L_2 L_1 = I$ to ensure continuity of the map along the l_n-axis. (This is constraint (3) above, which prevents us from ripping the paper.)

Theorem 3.1. *Given a foldable single-vertex crease pattern with associated folding map as defined above,* $\chi_1...\chi_{n-1}\chi_n = I$.

Proof. Using our notation from above and substituting our earlier definitions, we see that $L_n...L_1$

$$= (\chi_1...\chi_{n-1}\chi_n\chi_{n-1}^{-1}...\chi_1^{-1})(\chi_1...\chi_{n-2}\chi_{n-1}\chi_{n-2}^{-1}\cdot\chi_1^{-1})...(\chi_1\chi_2\chi_1^{-1})\chi_1$$
$$= \chi_1...\chi_{n-1}\chi_n.$$
This concludes the proof. $\square$

That $\chi_1...\chi_{n-1}\chi_n = I$ was stated without proof by Kawasaki in [7]. After we worked through the details, it became apparent to us that Kawasaki must have observed much of what we present here. We should note that Kawasaki himself states this criterion in the form of a definition: A crease pattern is foldable if and only if the condition $\chi_1...\chi_{n-1}\chi_n = I$ holds. (In contrast, recall our earlier definition of *foldable*.)

The converse of this statement is not true—there exist crease patterns such that $\chi_1...\chi_{n-1}\chi_n = I$, but which are not physically foldable.

Example: Examine the crease pattern $l_1 = (0, \pi/2), l_2 = (\pi/2, \pi), l_3 = (\pi, \pi/2), l_4 = (5\pi/4, -\pi), l_5 = (7\pi/4, \pi)$. The "flap" intersects one of the "walls." (See Figure 2.)

A special case of Theorem 3.1 is the previously known analogous result for single-vertex *flat* folds. Let $\Delta\alpha_i = \alpha_{i+1} - \alpha_i$ denote the angle between crease lines l_{i+1} and l_i. Given a single-vertex flat-fold, the folding angles ρ_i will all be $\pm\pi$ and the rotation matrices χ_i will be reflections over the crease lines in the xy-plane. Because the composition of two reflections is a rotation by twice the angle between the reflection lines, and we can think of the identity

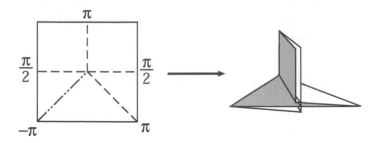

Figure 2. A fold that satisfies Theorem 3.1 but also self-intersects.

transformation as rotating about the origin by 2π, Theorem 3.1 becomes

$$\Delta\alpha_1 + \Delta\alpha_3 + \cdots + \Delta\alpha_{2k-1} = \Delta\alpha_2 + \Delta\alpha_4 + \cdots + \Delta\alpha_{2k} = \pi$$

where the crease pattern has $2k = n$ edges. (It is not hard to show that every single-vertex, flat-fold must have an even number of creases.) This is sometimes stated as $\Delta\alpha_1 - \Delta\alpha_2 + \Delta\alpha_3 - \cdots - \Delta\alpha_{2n} = 0$, and it is a necessary and sufficient condition for a single-vertex, fold to fold-flat. This was stated without proof by Justin [5] and Kawasaki [6] and first proven in [4].

4 Necessity for Multivertex Crease Patterns

The discussion of multivertex crease patterns requires some new notation. Here, we only consider bounded paper; in theory, there is no problem with infinite paper as long as the folding map is defined everywhere in the plane.

Consider our crease pattern with vertices, edges, and faces, and choose some vertex to lie at the origin. Label this vertex v_0. Enumerate the other vertices in some fashion as v_i, so that there are vertices $v_0, ..., v_b$. Some crease lines may be incident to the boundary of the paper; the points of incidence are "virtual vertices," since for indexing purposes we will want to label them, but they do not act as vertices for the purposes of our results. Denote these points as $(vv)_i$, and enumerate them $(vv)_{b+1}, ..., (vv)_c$. Now, denote the crease lines as $l_{(i_1,j_1)}$ where i_1, j_1 are the indices corresponding to the incident vertices (one of which may be virtual). Choose one face adjacent to v_0 to be fixed in the xy-plane, and denote it F_0. Finally, denote the other faces by $F_{(i_1,j_1,...)}$, where the coordinates of $(i_1, j_1, ...)$ are the indices corresponding to the vertices surrounding each face.

If we were to extend the analogy of the spider, then we would send the spider scurrying around the paper to define a map on each face. In order to describe the action of folding along crease line $l_{(i,j)}$, we want to describe an affine rotation about the line in which $l_{(i,j)}$ lies; thus, we will need to move to homogeneous coordinates. As before, each crease line $l_{(i,j)}$ is marked with an ordered pair $(\alpha_{(i,j)}, \rho_{(i,j)})$ corresponding to the angle to the x-axis and the folding angle. We will accomplish the crease by moving a vertex of $l_{(i,j)}$ to the origin, creasing as before, and returning $l_{(i,j)}$ to its original position. Let B_k be the 4×4 matrix which translates a point in $\mathbb{R}^3$ by v_k; let $A_{(i,j)}$ be the matrix in homogeneous coordinates which rotates the xy-plane by angle $\alpha_{(i,j)}$, and let $C_{(i,j)}$ be the matrix in homogeneous coordinates which rotates by angle $\rho_{(i,j)}$ in the yz-plane. Then the folding matrix will be $\chi_{((i,j),k)} = B_k A_{(i,j)} C_{(i,j)} A_{(i,j)}^{-1} B_k^{-1}$. Note that k indicates from which vertex we consider $l_{(i,j)}$ to extend, so that $k = i$ or $k = j$.

Eventually, we will define the folding map using these matrices.

Lemma 4.1. *Let v_k be any vertex in a foldable multivertex crease pattern, surrounded by crease lines $l_{(k,j_t)}$ and faces $F_{(k,j_t,\ldots)}$. If $F_0 = F_{(k,j_n,\ldots)}$ is fixed in the xy-plane, then*

$$\chi_{((k,j_1),k)} \cdots \chi_{((k,j_t),k)} \cdots \chi_{((k,j_n),k)} = I.$$

Proof. Because the order in which the matrices are applied is the same as in Theorem 3.1, the cancellation is also exactly the same, and we obtain the same result for the vertex v_k that we saw for a vertex at the origin. $\square$

Lemma 4.2. *Given two adjacent vertices v_i and v_j in a foldable crease pattern, suppose we have a curve γ that crosses crease line $l_{(i,j)} = (\alpha_{(i,j)}, \rho_{(i,j)})$, and $\chi_{((i,j),i)}$ is the matrix that rotates counterclockwise by $\rho_{(i,j)}$ about the axis corresponding to $l_{(i,j)}$. Also suppose that we have another curve γ' that crosses $l_{(i,j)}$ in the other direction, and let $\chi'_{((i,j),k)}$ be the matrix corresponding to $l_{(i,j)}$ from this other direction. Then $\chi'_{((i,j),i)} = \chi'_{((i,j),j)} = \chi_{((i,j),i)}^{-1}$.*

This lemma allows us to translate between matrices associated with curves crossing the same crease line, but in different directions or associated with different vertices.

Proof. Write

$$\chi_{((i,j),i)} = B_i A_{(i,j)} C_{(i,j)} A_{(i,j)}^{-1} B_i^{-1}$$

and

$$\chi'_{((i,j),k)} = B_k' A_{(i,j)}' C_{(i,j)}' (A_{(i,j)}')^{-1} (B_k')^{-1}.$$

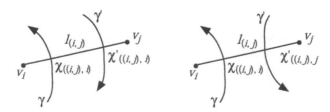

Figure 3. Curves crossing a crease line in different directions.

Let $\alpha_{(i,j)}$ and $\rho_{(i,j)}$ be the plane position and folding angles of $\chi_{((i,j),i)}$, respectively, and let $\alpha'_{(i,j)}$ and $\rho'_{(i,j)}$ be the corresponding angles for $\chi'_{((i,j),k)}$. Recall that $\alpha_{(i,j)}$ and $\alpha'_{(i,j)}$ will determine the matrices $A_{(i,j)}$ and $A'_{(i,j)}$, and that $\rho_{(i,j)}$ and $\rho'_{(i,j)}$ will determine the matrices $C_{(i,j)}$ and $C'_{(i,j)}$, respectively.

$\chi_{(i,j),k)}$ may be associated with either v_i or v_j (so $k = i$ or $k = j$); thus, we have two cases.

Case 1: $k = i$.
Then $\alpha'_{(i,j)} = \alpha_{(i,j)}$, and so $A_{(i,j)} = A'_{(i,j)}$. Also, $\rho'_{(i,j)} = -\rho_{(i,j)}$; because γ' is locally traveling clockwise around v_i, the rotation is in a direction opposite to our convention. So we have $C'_{(i,j)} = C^{-1}_{(i,j)}$, and thus

$$\chi_{((i,j),i)}\chi'_{((i,j),i)} = B_iA_{(i,j)}C_{(i,j)}A^{-1}_{(i,j)}B_i^{-1}B_iA_{(i,j)}C'_{(i,j)}A^{-1}_{(i,j)}B_i^{-1} = I.$$

Case 2: $k = j$.
As one can see in Figure 4, we have $\alpha'_{(i,j)} = \pi + \alpha_{(i,j)}$. This means that $A^{-1}_{(i,j)}A'_{(i,j)}$ corresponds to rotating the xy-plane by π. Denote this half-turn

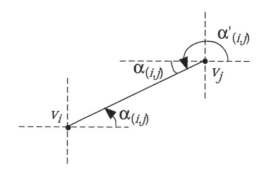

Figure 4. When χ and χ' are associated with different vertices.

by H. Then $A'_{(i,j)} = A_{(i,j)}H$. We also have $\rho'_{(i,j)} = \rho_{(i,j)}$ because γ' is now traveling counterclockwise around v_j. Thus $C_{(i,j)} = C'_{(i,j)}$. We may already write $\chi_{((i,j),i)}\chi'_{((i,j),j)} =$

$$B_i A_{(i,j)} C_{(i,j)} A^{-1}_{(i,j)} B^{-1}_i B'_j A_{(i,j)} H C_{(i,j)} H^{-1} (A_{(i,j)})^{-1} (B'_j)^{-1}.$$

For simplicity, let $\alpha = \alpha_{(i,j)}$ and $\rho = \rho_{(i,j)}$. Denote the coordinates $v_i = (x_1, y_1)$ and $v_j = (x_2, y_2)$. Then

$$B^{-1}_i B'_j = \begin{pmatrix} 1 & 0 & 0 & x_2 - x_1 \\ 0 & 1 & 0 & y_2 - y_1 \\ 0 & 0 & 1 & 0 \\ 0 & 0 & 0 & 1 \end{pmatrix}.$$

It can now be readily checked using a symbolic algebra program that if we let $S = (y_2 - y_1)/(x_2 - x_1)$, then $\chi_{((i,j),i)}\chi'_{((i,j),j)}$ can be written as

$$\begin{pmatrix} 1 & 0 & 0 & (x_2 - x_1)\sin\alpha\,(-sin\alpha + \sin\alpha\cos\rho + S\cos\alpha - S\cos\alpha\cos\rho) \\ 0 & 1 & 0 & (x_2 - x_1)\cos\alpha\,(\sin\alpha - \sin\alpha\cos\rho - S\cos\alpha + S\cos\alpha\cos\rho) \\ 0 & 0 & 1 & (x_2 - x_1)\,(-\sin\alpha\sin\rho + S\cos\alpha\sin\rho) \\ 0 & 0 & 0 & 1 \end{pmatrix}.$$

It is not immediately clear that this matrix is equal to the identity. But in this case, the slope of the crease line $l_{(i,j)}$ is $S = (y_2 - y_1)/(x_2 - x_1)$ and the angle it makes to the horizontal is α; thus, $\tan\alpha = (y_2 - y_1)/(x_2 - x_1)$. Using this fact with the above matrix reduces it to the identity. □

Example (Two-vertex crease pattern): We would like to extend Theorem 3.1 to multivertex crease patterns. In order to give an intuitive hint as to the proof of upcoming Lemma 4.3, let us examine a closed path on a two-vertex crease pattern. Figure 5 shows such a foldable pattern. In addition to the labeled vertices v_0 and v_1, we have virtual vertices $(vv)_2...(vv)_9$, labeled counterclockwise beginning with the one connected to v_0 and marked $(0, \pi/2)$.

Now, consider a closed path γ which traverses, in order, $F_0 = F_{(0,2,3)}$, $l_{(0,3)}, F_{(0,3,4,1)}, l_{(1,4)}, F_{(1,4,5)}, l_{(1,5)}, F_{(1,5,6)}, l_{(1,6)}, F_{(1,6,7)}, l_{(1,7)}, F_{(0,1,7,8)},$ $l_{(0,8)}, F_{(0,8,9)}, l_{(0,9)}, F_{(0,2,9)}, l_{(0,2)}, F_0$. Fixing F_0 and examining the matrices which correspond to this path, we have

$$\left(\chi_{((0,3),0)}\chi_{((1,4),1)}\cdots\chi_{((1,7),1)}\chi_{((0,8),0)}\chi_{((0,9),0)}\chi_{((0,2),0)}\chi^{-1}_{((0,9),0)}\cdots\chi^{-1}_{((0,3),0)}\right)$$
$$\left(\chi_{((0,3),0)}\cdots\chi_{((0,9),0)}\chi^{-1}_{((0,8),0)}\cdots\chi^{-1}_{((0,3),0)}\right)\cdots\left(\chi_{((0,3),0)}\chi_{((1,4),1)}\chi^{-1}_{((0,3),0)}\right)\chi_{((0,3),0)}$$
$$= \chi_{((0,3),0)}\chi_{((1,4),1)}\chi_{((1,5),1)}\chi_{((1,6),1)}\chi_{((1,7),1)}\chi_{((0,8),0)}\chi_{((0,9),0)}\chi_{((0,2),0)}.$$

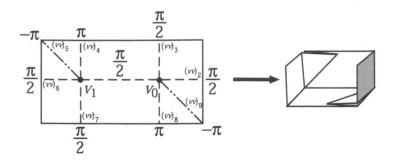

Figure 5. A nonflat crease pattern with more than one vertex.

By Lemma 4.1, we know that $\chi_{((0,3),0)}\chi_{((0,1),0)}\chi_{((0,8),0)}\chi_{((0,9),0)}\chi_{((0,2),0)} = I$. Then, by Lemma 4.2, $\chi_{((0,1),0)} = \chi_{((0,1),1)}^{-1}$. We may adapt Lemma 4.1 to show that

$$\chi_{((1,4),1)}\chi_{((1,5),1)}\chi_{((1,6),1)}\chi_{((1,7),1)}\chi_{((0,1),1)} = I$$

as well, so that

$$\chi_{((0,1),1)}^{-1} = \chi_{((1,4),1)}\chi_{((1,5),1)}\chi_{((1,6),1)}\chi_{((1,7),1)} \quad \text{and}$$

$$\chi_{((0,3),0)}\chi_{((1,4),1)}\chi_{((1,5),1)}\chi_{((1,6),1)}\chi_{((1,7),1)}\chi_{((0,8),0)}\chi_{((0,9),0)}\chi_{((0,2),0)} = I.$$

Lemma 4.3. *Consider a foldable multivertex crease pattern. Let γ be a closed curve which does not intersect any vertices, and which begins and ends on a face fixed in the xy-plane. Let $l_{(i_1,j_1)}, l_{(i_2,j_2)}, \dots l_{(i_n,j_n)}$ be the creases, in order, that γ crosses, and let $\chi_{((i_t,j_t),k_t)}$ be the rotation counterclockwise around $l_{(i_t,j_t)}$. Then*

$$\chi_{((i_1,j_1),k_1)}\chi_{((i_2,j_2),k_2)}\cdots\chi_{((i_n,i_n),k_n)} = I.$$

Proof. We proceed, agonizingly, by induction on the number of vertices enclosed by γ. The base case where γ encloses one vertex is completed in Lemma 4.1. Now suppose that γ encloses two or more vertices, and construct γ' by avoiding exactly one vertex v_y enclosed by γ (see Figure 6).

Let us denote $\Psi_t = \chi_{((i_1,j_1),k_1)}\chi_{((i_2,j_2),k_2)}\cdots\chi_{((i_t,j_t),k_t)}$. Then the complete sequence of matrices corresponding to γ is

$$(\Psi_{n-1}\chi_{((i_n,j_n),k_n)}\Psi_{n-1}^{-1})\cdots(\Psi_2\chi_{((i_3,j_3),k_3)}\Psi_2^{-1})(\Psi_1\chi_{((i_2,j_2),k_2)}\Psi_1^{-1})\chi_{((i_1,j_1),k_1)}$$

$$= \chi_{((i_1,j_1),k_1)}\chi_{((i_2,j_2),k_2)}\cdots\chi_{((i_n,i_n),k_n)}. \qquad (\star)$$

Case 1: v_y is not connected to any other vertex enclosed by γ. This means that the curve γ crosses all of the crease lines adjacent to v_y and that γ' crosses

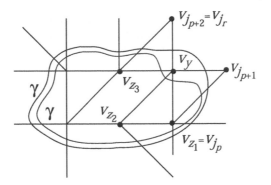

Figure 6. Inducting on the number of vertices enclosed by γ.

none of them. In particular, this subsequence begins and ends on the same face $F_{(j_r, j_p, \ldots)}$. Consider, within the sequence of matrices associated with γ, the subsequence Ψ which is comprised of all matrices corresponding to crease lines crossed to reach $F_{(j_r, j_p, \ldots)}$.

Now, the complete sequence $(\star)$ of matrices associated with γ will contain a subsequence equivalent, by Lemma 4.2, to the matrices associated to all of the crease lines adjacent to v_y, which we will denote $\chi_{(y, j_p), y)}$, $\chi_{(y, j_{p+1}), y)}$, $\ldots$, $\chi_{(y, j_{p+m}), y)}$ (in path order). We claim that the product of these matrices, as they appear in $(\star)$, $\chi_{(y, j_{p+m}), y)} \cdots \chi_{(y, j_p), y)}$ is equal to the identity. To see this, consider the complete sequence of matrices that correspond to traveling along γ to face $F_{(j_r, j_p, \ldots)}$ and then traveling along γ around v_y:

$$\left(\Psi \chi_{((y, j_p), y)} \cdots \chi_{((y, j_{p+m}), y)} \chi_{((y, j_p), y)}^{-1} \Psi^{-1} \right)$$

$$\cdots \left(\Psi \chi_{((y, j_p), y)} \chi_{((y, j_{p+1}), y)} \chi_{((y, j_p), y)}^{-1} \Psi^{-1} \right) \left(\Psi \chi_{((y, j_p), y)} \Psi^{-1} \right) \Psi$$

$$= \Psi \chi_{((y, j_p), y)} \cdots \chi_{((y, j_{p+m}), y)}.$$

Wherever face $F_{(j_r, j_p, \ldots)}$ is in space when we approach vertex v_y along γ, we can move it back to the xy-plane before approaching v_y to obtain

$$\Psi \chi_{((y, j_p), y)} \cdots \chi_{((y, j_{p+m}), y)} \Psi^{-1}.$$

This sequence of matrices places $F_{(j_r, j_p, \ldots)}$ in the xy-plane, travels around v_y, and then puts it back. Thus Lemma 4.1 tells us that this product is the identity, or $\chi_{((y, j_p), y)} \cdots \chi_{((y, j_{p+m}), y)} = I$. Thus, if we collapse these matrices in $(\star)$, the remaining matrices in the complete sequence for γ will be the same as those for γ', which equal the identity by the induction hypothesis.

Case 2: v_y is connected to some vertices $v_{z_1}, \ldots, v_{z_w}$ enclosed by γ, by crease lines $l_{(y, z_i)}$.

As γ passes around vertex v_y, denote the crease lines it crosses by $l_{(i_p,j_p)}$, $l_{(y,j_{p+1})}$, $l_{(y,j_{p+2})}, ..., l_{(y,j_r)}$, $l_{(i_s,j_s)}$. In the same region, γ' crosses crease lines $l_{(i_p,j_p)}, l_{(y,z_1)}, ..., l_{(y,z_w)}, l_{(i_s,j_s)}$. The folding maps defined by γ and γ' are the same until they reach crease lines $l_{(y,j_{p+1})}$ and $l_{(y,z_1)}$, respectively; denote this product of matrices by ψ.

Suppose we write out the complete sequence of matrices associated with γ and the complete sequence of matrices associated with γ'. By Lemma 4.1, $\chi_{((y,j_{p+1}),y)} \cdots \chi_{((y,j_{r-1}),y)} \chi_{((y,j_r),y)} \chi'_{(((y,z_w),y)} \cdots \chi'_{((y,z_1),y)} = I$. Then, using Lemma 4.2 we see that $\chi'_{(((y,z_t),y)} = \chi^{-1}_{(((y,z_t),z_t)}$, and so

$$\chi_{((y,j_{p+1}),y)} \cdots \chi_{((y,i_{r-1}),y)} \chi_{((y,i_r),y)} = \chi_{(((y,z_1),z_1)} \cdots \chi_{((y,z_w),z_w)}.$$

Using these facts, the complete sequence of matrices associated with γ may be rewritten exactly as the sequence of matrices associated with γ', which, by the inductive hypothesis, equals I.   □

We would like to use a product of χ matrices to define our folding map. Before doing so, we must be sure it is well-defined:

Lemma 4.4. *The definition of the folding map on face $F_{(i_p,j_p,...)}$ is independent of the defining path γ chosen from the fixed face F_0 in the xy-plane.*

Sketch of Proof: Consider any two vertex-avoiding paths γ_1 and γ_2 from face F_0 to face $F_{(i_p,j_p,...)}$. Using the above lemmas and the fact that $\gamma_1\gamma_2^{-1}$ is a closed curve, it is straightforward to see that the product of matrices $\chi_{((i,j),k)}$ for these two paths are the same. (The interested reader may see [1] for more details.)   □

Thus, we may employ our helpful spider in the definition of our folding map; it need merely be intelligent enough to cross each crease line only once. Lemma 4.4 assures us that no matter which path the spider takes to reach a given face, the map it generates will be well-defined.

More specifically, we can define the general folding map f as follows: Let F_0 be the face that is fixed in the xy-plane. Given any other face $F_{(i_p,j_p,...)}$, let γ be any vertex-avoiding path from a point in F_0 to a point in $F_{(i_p,j_p,...)}$. Let the crease lines that γ crosses be, in order, $l_{(i_1,j_1)}, ..., l_{(i_p,j_p)}$. Then the folding map for $(x,y) \in F_{(i_p,j_p,...)}$ is,

$$f(x,y) = f(x,y,0,1) = \chi_{((i_1,j_1),i_1)} \chi_{((i_2,j_2),i_2)} \cdots \chi_{((i_p,j_p),i_p)}(x,y,0,1).$$

Theorem 4.1. *Consider a foldable multivertex crease pattern. Let γ be a closed curve, beginning and ending on face $F_{(i_p,j_p,...)}$, which does not intersect any vertices. Let $l_{(i_1,j_1)}, l_{(i_2,j_2)}, ... l_{(i_n,j_n)}$ be the creases, in order, that γ crosses, and let $\chi_{((i_t,j_t),k_t)}$ be the rotation counterclockwise around the crease line $l_{(i_t,j_t)}$. Then $\chi_{((i_1,j_1),k_1)} \chi_{((i_2,j_2),k_2)} \cdots \chi_{((i_n,j_n),k_n)} = I$.*

Proof. For simplicity, let us denote $\chi_{((i_1,j_1),k_1)}\chi_{((i_2,j_2),k_2)}\cdots\chi_{((i_t,j_t),k_t)}$ by Ψ_t. Using the definition of the folding map, let γ_1 be a vertex-avoiding path from a point in F_0 to a point in $F_{(i_p,j_p,\ldots)}$ and let $\varphi = f(F_{(i_p,j_p,\ldots)})$. Consider the path γ_2 equivalent to γ_1 followed by γ; by Lemma 4.4, γ_1 and γ_2 define the same folding map on $F_{(i_p,j_p,\ldots)}$. Thus,

$$\Psi_{n-1}\chi_{((i_n,j_n),k_n)}\Psi_{n-1}^{-1}\cdots\Psi_1\chi_{((i_1,j_1),k_1)}\Psi_1^{-1}\chi_{((i_1,j_1),k_1)}$$

$$= \chi_{((i_1,j_1),k_1)}\chi_{((i_2,j_2),k_2)}\cdots\chi_{((i_n,j_n),k_n)}\varphi = \varphi$$

or

$$\chi_{((i_1,j_1),k_1)}\chi_{((i_2,j_2),k_2)}\cdots\chi_{((i_n,j_n),k_n)} = I.$$

$\square$

Corollary 4.1. *Let v_k be any vertex other than the origin in a foldable multivertex crease pattern, surrounded by crease lines $l_{(i_p,j_p)}$ and faces $F_{(i_p,j_p,\ldots)}$. Then*

$$\chi_{((k,j_1),k)}\chi_{((k,j_2),k)}\cdots\chi_{((k,j_n),k)} = I.$$

Acknowledgements

The authors were supported in this project by 2000 Summer Fellowship Grants from the University of Northern Iowa and Merrimack College.

References

[1]   s-m. belcastro, T. Hull, "Modelling the folding of paper into three dimensions using affine transformations", *Linear Algebra and Its Applications*, to appear.

[2]   M. Bern, B. Hayes, "The complexity of flat origamis", *Proceedings of the 7th Annual ACM-SIAM Symposium on Discrete Algorithms*, (1996) 175-183.

[3]   D. Fuchs, S. Tabachnikov, "More on paperfolding", *American Mathathematical Monthly*, Vol. 106, No. 1, (1999) 27-35.

[4]   T. Hull, "On the mathematics of flat origamis", *Congressus Numerantium*, Vol. 100, (1994) 215-224.

[5]   J. Justin, "Mathematics of origami, part 9", *British Origami*, (June 1986) 28-30.

[6]   T. Kawasaki, "On the relation between mountain-creases and valley-creases of a flat origami" (abridged English translation), in H. Huzita ed., *Proceedings of the First International Meeting of Origami Science and Technology*, (Ferrara, 1989) 229-237.

[7]   T. Kawasaki, "$R(\gamma) = I$", in K. Miura ed., *Origami Science and Art: Proceedings of the Second International Meeting of Origami Science and Scientific Origami*, (Seian University of Art and Design, Otsu, 1997) 31-40.

The Definition of
Iso-Area Folding

Jun Maekawa

1 The Definition of Iso-Area Folding

The concept of iso-area folding was presented by Kawasaki [1], who made iso-area cubes, an iso-area octahedron, and others. Figure 1 shows crease patterns of his iso-area cubes. These models are more highly symmetric than other origami models. To understand these models, symmetry analysis is useful.

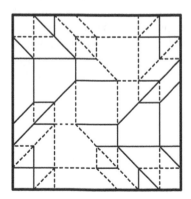
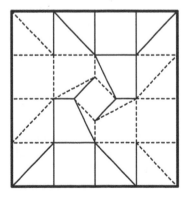

Figure 1. Crease patterns of Kawasaki's iso-area cubes.

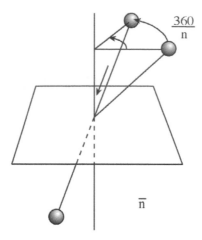

Figure 2. Rotation inverse operation.

The conclusion is very simple. Iso-area folds should be regarded as posessing a kind of "rotational inverse symmetry".

The combined operation of rotation by $360/n$ degrees (where n is a natural number) and inversion through the origin is called the "n-fold rotation inverse operation". The sequence of rotation and inversion can be changed. When a figure (solid) is not changed by an n-fold rotation inverse operation, it is said to have "n-fold rotational inverse symmetry," denoted by n-bar (or $\bar{n}$, see Figure 2). I will use "n-bar symmetry" instead of "n-fold rotational inverse symmetry" in this paper, for brevity and clarity.

The original concept of iso-area models had to do with the finished model, and thus it is natural to focus on what kind of n-bar symmetry the finished model has. However, this symmetry will also be present in the crease patterns of such models, where we assume the origin to be in the center of the paper and inversion through the origin will change all mountain creases to valleys and vice-versa (as well as transforming points (x, y) to $(-x, -y)$). With this realization, we notice that both of Kawasaki's cubes have 4-bar symmetry by examining their crease patterns in Figure 1.

Figure 3 shows the crease pattern of an iso-area octahedron skeleton by Maekawa. This octahedron skeleton has a slightly different character from Kawasaki's cubes, but it is also an iso-area fold. It has 1-bar symmetry (or just "inverse symmetry").

In general, we would like to define an origami model to be iso-area if it has n-bar symmetry for any natural number n. However, the case of $n - 2$ presents difficulties, because it is equivalent to mirror symmetry and is not iso-

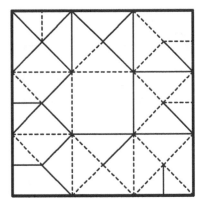

Figure 3. Crease pattern of iso-area octahedron skeleton 1 (Maekawa).

area folding. Figure 4 shows that the classic Japanese crane (which is clearly not iso-area) has 2-bar symmetry. Why does this happen?

The reason has to do with the relationship between the axis of rotation and the orientation of the paper. Look at the crane example again (Figure 4). The axis of rotation penetrates the model through the side, as opposed to through

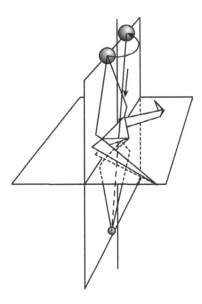

Figure 4. Mirror symmetry.

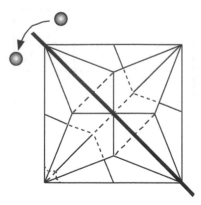

Figure 5. For the crane, the axis of rotation is along a diagonal (2-bar symmetry).

the top and bottom. Think of what this would look like on the unfolded piece of paper with the crease pattern drawn on it, as in Figure 5. The axis of rotation would go through a diagonal of the square! This rotation would flip the paper over, switching all the mountains to valleys and vice-versa. Then the inverse operation would put the creases back into their original places, and restore the mountain and valleys to their original orientation. In the iso-area models we have seen thus far, the axis of rotation has been perpendicular to the unfolded sheet of paper. The thing to avoid, then, is not necessarily 2-bar symmetry, but the situation where the axis of rotation is not perpendicular to the original sheet of paper.

Thus we may make the following definition: An origami model is *iso-area* if it has n-bar symmetry and the rotation axis passes through the plane of the unfolded paper perpendicularly. In the case of 2-bar symmetry, the model would have to be "no fold" origami (i.e., an uncreased square) in order to be iso-area.

2   Some Characteristics of Iso-Area Models

First, notice that replacing a mountain with a valley fold is not the inverse operation. The inverse operation is equivalent to replacing a mountain with a valley fold and rotating it 180 degrees (Figure 6).

As already mentioned, there are many examples of iso-area folding. Table 1 shows known and unpublished iso-area models. (Note that some of these models may have other types of symmetry as well. For example, the "Hexa Cube" has 1-bar symmetry as well as 3-bar.)

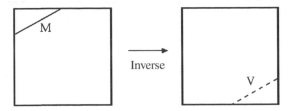

Figure 6. Inverse operation in origami.

The most interesting point of this table is that the Miura-ori (the famous space engineering result for folding a large sheet) is also iso-area folding. Miura-ori is one of the 1-bar symmetry models. one-bar symmetry forms are more flexible than others, in that when folded they can be opened up and closed again in a single motion. The MARS pattern (by Paulo Tabolta Barreto [2]) is a good example. Miura-ori is constructed by one type of parallelogram, but MARS p is constructed by two types of parallelograms (Figure 7). One of them (the square) has 4-fold rotational symmetry. This means that if we choose the mountains and valleys carefully, we can fold this pattern into 4-bar symmetry model. But if we do this, the result is not flexible at all; the surfaces of the parallelograms bend. Also, we cannot fold it by combining 4-bar symmetry creases only. It seems that 1-bar symmetry has particular meaning for transformable sheet structures.

	Model	Creator	Type	Year
1	Cube 1	Kawasaki	4-bar	1985?
2	Cube 2	Kawasaki	4-bar	1985?
3	Octahedron	Kawasaki	3-bar	1985?
4	Octahedron Skeleton 1	Maekawa	1-bar	1985
5	Octahedron Skeleton 2	Maekawa	4-bar	1985
6	Miura-ori	Miura	1-bar	197?
7	MARS	Barreto	1-bar	1994?
8	Tri-color Puzzle	Neal	1-bar	?
9	Flower Tower	Palmer	4-bar	1994?
10	Hexagonal Folds	Hull	3-bar	1998?
11	Cubes	Maekawa	1-bar	2000
12	Diamond Cube	Maekawa	1-bar	2000
13	Silver Cube	Maekawa	1-bar	2000
14	Hexa Cube	Maekawa	3-bar	2000

Table 1. Iso-area origami models.

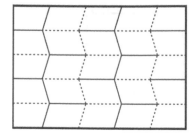
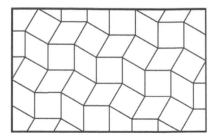

Figure 7. Miura-ori (Miura) and MARS (Barreto).

Finally, I introduce some new models. Figures 8, 9, and 10 show unfolded patterns of cubes. These models are based on symmetric cross-sections of the cube (Figure 11). Please try to fold them to see the relationship.

References

[1] Kasahara, K. and Takahama, T., *Origami for the connoisseur*, Japan Publications Inc., (1987).

[2] Miura, K. ed., *Origami Science and Art, Proceedings of the 2nd International Meeting of Origami Science and Scientific Origami*, Seian University, (1997).

[3] Nagakura, S., *Rikagaku Jiten (The Encyclopedia of Physics and Chemistry), 5th edition*, (1998).

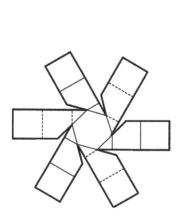
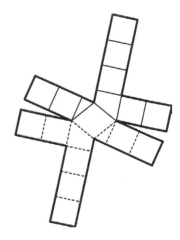

Figure 8. Hexa Cube (Maekawa).   **Figure 9.** Diamond Cube (Maekawa).

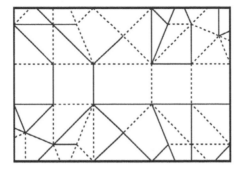

Figure 10. Silver Cube (Maekawa).

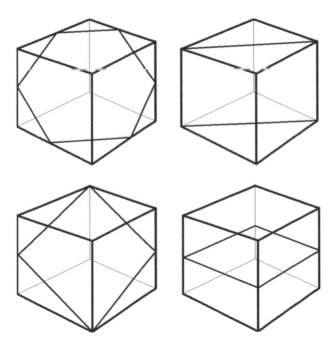

Figure 11. Symmetric planes of the cube.

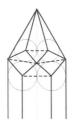

The Geometry of Orizuru

Toshikazu Kawasaki

1 Deformation of a Bird Base

Most origami is folded from colored square paper. Classic examples are the *yakko* (serving man) and the *orizuru* (crane). But it is possible to fold the orizuru from paper other than a square. If you fold along the creases made in rhombic paper as in Figure 1, you can make an *orizuru* with long wings. Many people have been aware of this fact from old times. However, you cannot make a beautiful *orizuru* from a rectangle.

Kodi Husimi [1] thought of folding an *orizuru* from kite-shaped paper (Figure 2) when he was working on a flying *orizuru*. The details will be explained later, but it is not simple to fold the *orizuru* from kite-shaped paper.

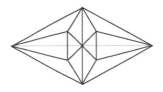

Figure 1. Crease pattern for Orizuru with long wings.

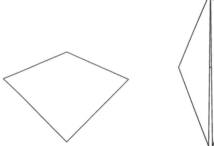

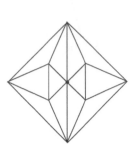

Figure 2. Orizura from   **Figure 3.** The bird base.   **Figure 4.** A development
kite-shaped paper.   chart of the bird base.

Figure 3 shows an origami model called the *bird base*, which is often
employed when making birds. The lower flaps are narrowed into slender forms,
and they become a neck and a tail (legs) when folded up between wings. Figure
4 is a development chart of the bird base. In the chart, the central point becomes
the center of the back of *orizuru*, and it is called the *bird base center*.

2   Neck-Wing Interchangeability

In the ordinary bird base, you can interchange the positions of the neck and tail
with the wings by moving the corners up and down (Figure 5(a)). We will call
this property *neck-wing interchangeability*. The exchange of the neck and tail
with wings corresponds to the change of fold lines as shown in development
charts (b)–(d).

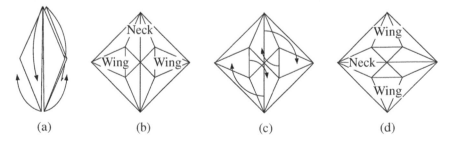

(a)   (b)   (c)   (d)

Figure 5. Neck-wing interchangeability.

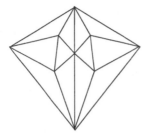

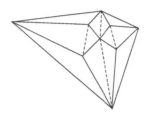

Figure 6. Husimi-zuru.　　　　　**Figure 7.** Justin-zuru.

3　Husimi-zuru and Justin-zuru

Husimi-zuru is a generalized *orizuru* made from kite-shaped paper of (Figure 6). Jacques Justin [3] advanced *Husimi-zuru* and for the first time worked out a generalized *orizuru* from a convex quadrilateral with an inscribed circle, which has a neck-wing interchangeability. This is called *Justin-zuru* (Figure 7).

4　Maekawa-zuru

The deformation of an *orizuru* is possible by changing the shape of the paper, but it is also possible for regular square paper by destroying the symmetry of fold lines. Jun Maekawa, who is known for his masterpiece "Devil" model and his geometrical origami designs, showed us this kind of deformation in his *"New Orizuru"* (see [4]). In this paper, we shall call it *Maekawa-zuru*.

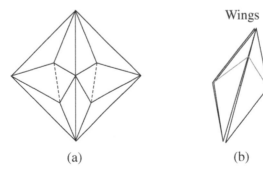

Wings

(a)　　　　　　　　　　(b)

Figure 8. *Maekawa-zuru.*

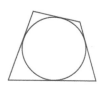 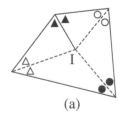 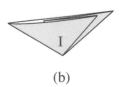

(a)   (b)

Figure 9. A good quad.   **Figure 10.** Bisection fold of a quadrilateral.

5 Good Quads

A circle circumscribed by a polygon is called an *inscribed circle* (Figure 9). This section deals with an interesting property of quadrilaterals with inscribed circles. We may bisect each angle of the quadrilateral by folding each angle in half. If these fold lines meet at a point, the quadrilateral can be folded flat along these fold lines.

Definition 6.1: When you fold a quadrilateral flat by bisectors of the four corners, this fold is called the *bisection fold of the quadrilateral* (Figure 10).

Unlike the case of a triangle, however, the bisection fold of the angles cannot be applied to all the quadrilaterals. It can be applied only to those quadrilaterals which have an inscribed circle. We call such quadrilaterals *good quads* in this paper. They are closely related to the deformation of *orizuru*.

Proposition 6.1: The necessary and sufficient condition for quadrilateral ABCD to be folded on the bisector of the angle is that it is a good quad. However, when the quadrilateral has a re-entrant angle, the inscribed circle and re-entrant sides must be tangent on their extension lines (Figure 11).

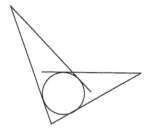

Figure 11. Inscribed circle and re-entrant side are tangent.

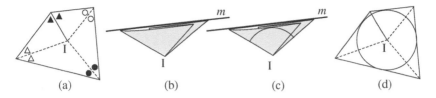

Figure 12. Proof of the necessary condition.

Proof of the necessary condition. According to Definition 6.1, fold lines or bisectors of the angle intersect at a point. We name this point I in Figure 12(a). The two sides bisected by a fold line overlap when folded, so the four sides of the quadrilateral overlap on the same straight line m when folded (Figure 12(b)). On the folded form, if we draw a circle around central point I such that it is tangent to m , then it is tangent to the four sides when opened. Namely, it is an inscribed circle. (Although Figure 12 does not illustrate it, this method also works for non-convex good quads.) □

Proof of the sufficient condition. If we draw perpendicular lines toward the four sides from the center I of the inscribed circle, we obtain four groups of triangles which have jointly oblique sides. The perpendicular lines are radii of the inscribed circle, so they are all equal in length. Accordingly, the triangles of each group are congruent, and each oblique side is a bisector. Moreover, in Figure 13(b), $\angle AIB + \angle CID = (a+b) + (c+d) = (a+d) + (b+c) = \angle AID + \angle BID$ and $2(a+b+c+d)= 360°$ and $\angle AIB + \angle CID = \angle AID + \angle BIC = 180°$. Therefore, the four lines IA, IB, IC, ID satisfy the locally flat-folding condition, and it is possible to fold quadrilateral ABCD on the bisector of the angle.

For sufficiency in the nonconvex case, see Figure 14. The crease lines are IA, IB, CI, and DI. Notice that $(p+r) - (q+s) = (a+f+c+2d+e) - (a+2b+c+e+f) = 2(d-b)$. Also, $\triangle CEO \sim \triangle CIO$ (so $2f + e = 2d + e$)

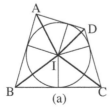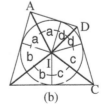

Figure 13. Proof of the sufficient condition.

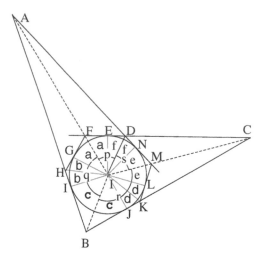

Figure 14. Suficiency of the nonconvex case.

and $\triangle ANO \sim \triangle AIO$ (so $2f + a = 2b + a$). Thus $d = f = b$, and we get $(p + r) - (q + s) = 0$, which means the crease lines satisfy the locally flat-folding condition. $\square$

The following proposition is well-known:

Proposition 6.2: A quadrilateral ABCD has an inscribed circle if and only if AB + CD = BC + DA.

This proposition is strange. Even if we determine the lengths of the sides of a quadrilateral, the shape is undecided. Nonetheless, if it has an inscribed circle in certain conditions, the proposition guarantees that it will not lose the property of having an inscribed circle no matter how it is deformed (Figure 15).

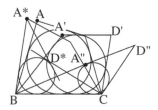

Figure 15. A quadrilateral with an inscribed circle.

In Figure 15, three vertices B, A″ and D″ lie on a straight line. Quadrilateral A″BCD″ is no longer a quadrilateral, but a triangle. Quadrilateral A*BCD* is re-entrant, but in the process of transformation from the original quadrilateral ABCD, it becomes a triangle once. How should we deal with a triangle like this? Mathematics allows no exceptions.

Definition 6.2: Consideration of a triangle as a quadrilateral by regarding one of the contact points of the inscribed circle as a vertex will be defined as a *quadrilateralization*. This contact point will be called a *flat vertex*. Such triangles are called *good flat quads* as all the convex and re-entrant quadrilaterals which have a quadrilateralized triangle and an inscribed circle are called *good quads*.

To join quadrilaterals of which two sides match is called a *two-sided union*. With regard to such joinings, the good quad has a "good" property.

Proposition 6.3: The two-sided union of good quads will make another good quad.

Proof. In Figure 16(a), by Proposition 6.2, $a+q = b+p$ and $c+p = d+q$. The sum of the two equations gives $a + c = b + d$. Therefore, it has an inscribed circle by Proposition 6.2.   □

If you apply Proposition 6.3 to a triangle, you will obtain the following:

Proposition 6.4: Given two triangles with a common side, if the feet of the perpendiculars drawn from the in-centers of the triangles toward the common side are identical, then the quadrilateral made by joining the two triangles has an inscribed circle (Figure 17).

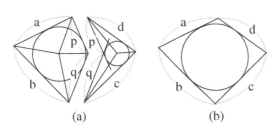

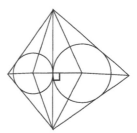

Figure 16. The two-sided union of good quads.   **Figure 17.**

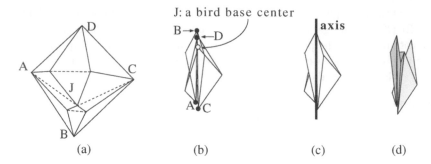

Figure 18. A generalized bird base. (c) Viewed from the side.

6   Generalized Bird Bases

Since the foundation has been laid, we now go on to the main subject.

Definition 7.1: When it is possible to fold a quadrilateral ABCD flat—for example on the creases as shown in Figure 18(a), and the four sides of the quadrilateral are, after folding, overlapped on the straight line which runs through point J, Figure 18(b) will be called a *generalized bird base*, and point J the *(generalized) bird base center* and the straight line (Figure 18(c)) the *axis (of the generalized bird)*.

Figure 19 illustrates a piece of paper being folded in half, cut straight and opened. The slits are symmetrical with respect to the fold line.

Figure 20(a) is the generalized bird base from Figure 18 that has been cut along the axis and opened. It is separated into four parts, P_1, P_2, P_3 and P_4. Figure 20(b) shows those parts joined together. In the same way as Figure 19, the slits are symmetrical with respect to the fold lines. For instance, in part P_1 in Figure 20(c), B′B and B′J are symmetrical with respect to fold line B′I$_1$. This means that fold line B′I$_1$ is the bisector of angle B′. Other fold lines

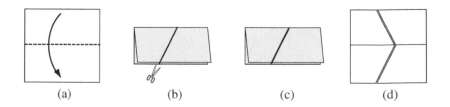

Figure 19.

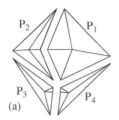
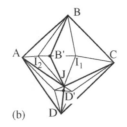
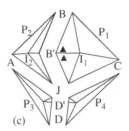

Figure 20.

and parts are the same. As a result, we obtain the following proposition using Proposition 6.3.

Proposition 7.1: If you cut a generalized bird base along the axis, it is separated into four good quads, each having the bisection fold of the angle property.

Quadrilateral AJCB of Figure 20(b) is a two-sided union of P_1 and P_2, which are good quads, so it also becomes a good quad according to Proposition 4.4. In the same way, quadrilateral ADCJ is a good quad. Since it is possible to make a two-sided union with quadrilateral AJCB and quadrilateral ADCJ, the resultant quadrilateral ABCD also becomes a good quad. Thus, we obtain the following proposition:

Proposition 7.2: A paper shape from which we can fold a generalized bird base must be a good quad.

It has now become clear that we cannot fold the generalized bird base from paper other than a good quad. Conversely, can we fold generalized birds if the paper shape is a good quad? It is simple to draw a good quad. Draw four lines in a circle so that they are circumscribed, and try to fold the generalized bird from the paper.

7   Open Good Quads

If we fold a sheet of paper flat on the four fold lines which satisfy the locally flat-folding condition, cut straight, and open, we have a good quad. In this way, we can easily make a quadrilateral for folding a generalized *orizuru*. If we cut the paper in a different way, the good quad changes accordingly. The methods of folding a generalized *orizuru* from the convex good quad (kite-shape quadrilateral) S of Figure 21(d) and the flat good quad (triangle) T have already been introduced in Section 5. If we bring cut S close to cut R, the

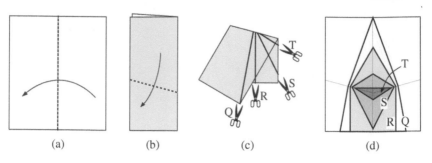

Figure 21.

good quad becomes oblong, and the moment when they meet, the lower end opens. The same applies to cut Q.

Open quadrilaterals Q and R have an inscribed circle as shown in Figure 22. Although they are open, we would like to regard them as good quads. Can we fold a generalized *orizuru* from them?

Taking the kite-shape quadrilateral as a model, let's fold the *orizuru* from R (Figure 23).

(1) Divide the quadrilateral by an axis of symmetry and find the *orizuru* center (?) by folding one part on the bisector of the angle.

(2) Link this point and the four vertices and divide it into four triangles.

(3) & (4) Fold each triangle on the bisector of the angle.

We can actually fold generalized *orizurus* on these creases; from (3), the *orizuru* which has symmetric wings and an infinite tail, and from (4), the *orizuru* which is symmetric in front and behind and one of the wings is infinite.

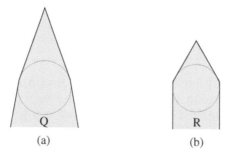

Figure 22.

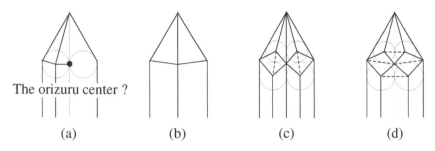

Figure 23.

If we interpret the good quad broadly as above, we can make strange generalized *orizurus*. The examples in Figure 24 do not look like quadrilaterals apparently, but they become quadrilaterals when segmented at the vertices marked •. Also, not all the inscribed circles seem to contact the four sides. In Figure 24(a), however, the inscribed circle contacts sides a and b at point P, and sides c and d at point Q, so (a) is a good quad. In the same way, the other quadrilaterals are good quads.

If you feel interested, try to fold generalized *orizurus* from these open good quads. First, try to fold open kite-shape paper, making use of the above method (Figure 23) of marking the *orizuru* center.

8 Reconsideration of Good Quads

Figure 25 shows the variations of an open good quad (Figure 24(a)). If the angle of the side increases by more than 180 degrees, it becomes a reentrantquadrilatera with an inscribed circle as shown in (d), but it is more natural to regard it as shown in (e).

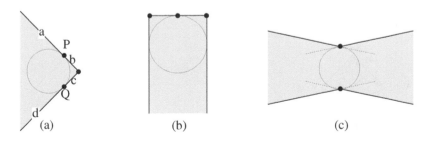

Figure 24.

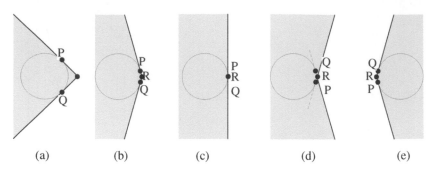

Figure 25. Variations of an open good quad.

Figure 25(e) is identical to Figure 25(b). This means that we can fold a bird base on the opposite side (outside). If we develop this idea further, we can also fold a generalized bird on the outside of a closed good quad as shown in Figure 26.

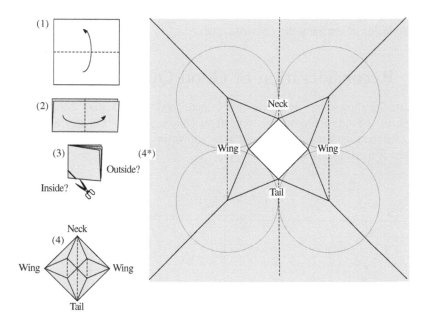

Figure 26.

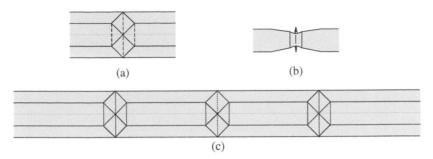

(a)   (b)

(c)

Figure 27.

9   Applications of a Generalized *Orizuru*

Lastly, I introduce three application examples of a generalized *orizuru*. Figure 27(b) is a generalized *orizuru* of a good quad, of which two parts are open, as shown in Figure 27(a). Since the neck and tail of the *orizuru* is extremely short, it looks like a turtle. We may call it *"turtle-crane"*. (In Japan, the crane and turtle are symbols of long life and happiness, so it is a doubly auspicious crane.) The turtle-crane can be linked as in (c).

Appendix

This paper is a condensation of [6], to be published, so you will be able to learn more about deformation of *orizuru* and other geometric topics of origami.

References

[1]   Husimi, K and Husimi, M., *The Geometry of Origaimi*, Nihon Hyoron-sha (1979).

[2]   Huzita, H., ed., *Origami Science and Technology, Proceedings of the First International Meeting of Origami Science and Technology* (1990).

[3]   Justin, J., "Mathematical Remarks and Origami Bases", *Symmetry: Culture and Science*, Vol.5, No.2 (1994), 153-165.

[4]   Kasahara, K. and Maekawa, J., *Viva Origami*, Sanrio (1983).

[5]   Kawasaki, T., "Expansions and Their Applications of Systematic Compositions of Cell Decompostions of Flat Origami, - Theory of Deformations of Orizuru", *Research Report of Sasebo College of Technology*, No. 32 (1995), 29-58.

[6]   Kawasaki, T., *Rose, Origami and Mathematics*, Japan Publications (To appear).

The Validity of the Orb, an Origami Model

Jeannine Mosely

It is possible to build apparent physical representations of impossible geometry, because our materials and our methods of preparing them are not perfect, especially when the medium is paper. But the validity or lack of validity of a model need not interfere with our aesthetic enjoyment of it, though it certainly prevents us from using it as an example of an object that exhibits certain theoretical properties. As a mathematician and an origamist, I am naturally curious about the validity of many origami models that I encounter.

For example, let us consider an origami model described by Dave Mitchell in his book *Mathematical Origami* [2]. His "ring of five cubes" is shown in Figure 1. This model is assembled from five cubes that have each had a

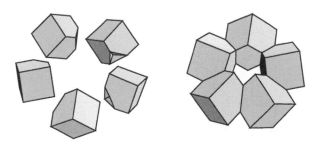

Figure 1. Mitchell's ring of five cubes.

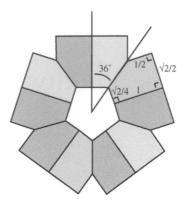

Figure 2. Overhead view of the ring of five cubes.

corner "dimpled in". The dimples are made by folding along lines connecting the midpoints of each of the three edges that meet at the corner. The ring is formed by inserting one corner of each cube into the dimple of its neighbor. This model is easy to build, even though it turns out to be invalid.

To prove this, we appeal to the symmetry of the model. Examining the overhead view of the ring of five cubes in Figure 2 we see, by symmetry, that the two lines indicated must be at an angle of $\pi/5$, or 36 degrees. However, if we consider the construction method of the dimpled cubes and the requirement that ideal paper may neither stretch nor compress, we can also infer the projected lengths of the various edges as marked. With a little bit of trigonometry, the edge lengths imply that this angle must be $\tan^{-1}(\sqrt{2}/2)$ or 35.26 degrees. Hence the model is not valid.

Let us now consider the model called the Orb [3], which I discovered in 1997 while investigating origami with curved creases. It is shown in Figure 3,

 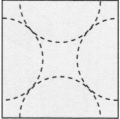

Figure 3. The Orb and its crease pattern.

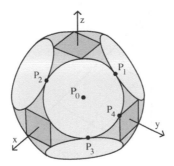

Figure 4. A simplified view of the Orb.

along with its crease pattern. The Orb is made from six identical modules, each folded from a square with four semicircular creases centered on the midpoints of each of the square's four edges, where each circle is tangent to both of its neighbors. It will prove convenient to let the radius of the circles be 1. The diagram shows the Orb as it would appear made from three different colors or shades of paper. It is not immediately obvious whether this model is valid.

A na•ve interpretation of the Orb is to treat it as if it were made of eight inverted cones connected by some planar and generalized cylindrical surfaces. Figure 4 shows a simplified view of the Orb with the cones replaced by circular faces. The Orb is centered at the origin, and oriented so that the axes pass through the centers of the square planar regions. In this picture, three different shades are used to distinguish the different types of surfaces, and do not reflect the coloring of an actual model.

By symmetry arguments, we see that the points P_1, P_2, and P_3 must have the form $(0, k, k)$, $(k, 0, k)$, and $(k, k, 0)$, where k is some unknown constant. Furthermore, the center of the base of the cone is the barycenter of P_1, P_2, P_3, so

$$P_0 = (P_1 + P_2 + P_3)/3 = (2k/3, 2k/3, 2k/3).$$

The radius of the base circle can be determined by observing that the cones are made up of three quarter circles. Since the radius of those circles is 1, the circumference of the base is $3\pi/2$, and therefore the radius of the base is $3/4$. This allows us to solve for k, and we find that $k = 3\sqrt{6}/8$. It is easy to show, by some vector algebra, that the point P_4, opposite P_2 on the circle, has the form

$$P_4 = (k/3, 4k/3, k/3) = (\sqrt{6}/8, \sqrt{6}/2, \sqrt{6}/8).$$

Now let us refer to Figure 5, which shows the portion of the paper, prior to folding, that is bounded by P_4, P_0, P_3 and the center of the square face pierced

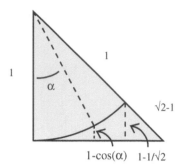

Figure 5. A fundamental region of the Orb's crease pattern.

by the y-axis in Figure 4. We see that height of P_4 above the xy-plane must be $1 - 1/\sqrt{2} = .293$ which is not equal to the z-coordinate of P_4, $\sqrt{6}/8 = .306$.

At this point, we might supppose that the Orb is an invalid model. It turns out, however, that we made an unwarranted assumption when we concluded that the base of the cone lies in a plane, and our analysis of its geometry was incorrect. The base almost lies in a plane, and it is hard to see that it does not. Physical models can deceive us. By existing, they make us believe in them, and because our vision is not perfect, we look at them and misinterpret their shape.

We now re-analyze the Orb by making some symmetry arguments about it and setting up a system of constraints on its geometry that are implied by its construction. Note that the Orb can be viewed as a cube that has had its corners dimpled in, but instead of being dimpled in along straight lines, it is dimpled in along curved creases. In Figure 6, we dissect the cube and the

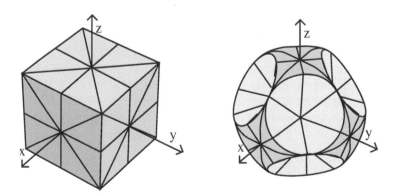

Figure 6. Fundamental regions of the cube and the Orb.

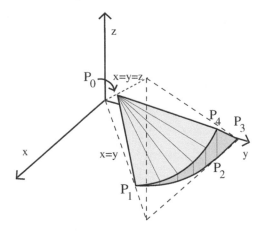

Figure 7. A single fundamental region of the Orb.

Orb into their fundamental regions under the set of symmetry operations on the cube [1]; that is, we slice them into the smallest pieces such that every piece can be transformed into every other by a suitable combination of rotations and reflections.

The fundamental region of the cube is a tetrahedron whose base is a right isoceles triangle and whose fourth vertex is at the center of the cube. Appropriately rotated and reflected, 48 similar tetrahedra combine to make the cube. The fundamental region for the Orb is shown in Figure 7. It must fit snugly inside this tetrahedron in order for the 48 rotated and reflected regions to join seamlessly to form the Orb. That is, P_0 must lie on the line $x = y = z$, P_1 must lie on the line $x = y$, $z = 0$ and P_3 must lie on the y axis. In addition, the curved line connecting P_1 and P_2 must lie in the xy-plane and the right isoceles triangle $\triangle P_2 P_3 P_4$ must be parallel to the xz-plane.

These are merely the constraints imposed by symmetry. To complete our analysis, we must also consider the constraints imposed by the Orb's construction. Clearly, $P_2 P_3 = P_2 P_4 = 1 - 1/\sqrt{2}$ and $P_3 P_4 = \sqrt{2} - 1$. Let $c = (c_x(\alpha), c_y(\alpha), c_z(\alpha))$ be the curve connecting P_1 and P_4, parameterized with respect to arc length α measured from P_1. Then any point $c(\alpha)$ on c must be a distance 1 from P_0. Furthermore, from Figure 5 it is evident that the z-coordinate of a point on the curve $c_z(\alpha) = 1 - cos(\alpha)$ and that the range of α is $[0, \pi/4]$. Finally, $P_0 = (h, h, h)$ for some value of h.

So, to demonstrate the Orb's validity, we must find functions $c_x(\alpha)$, $c_y(\alpha)$, $c_z(\alpha)$ and the value of h for which all these constraints hold.

The expressions defining $c(\alpha)$ are greatly simplified by an appropriate change of coordinates. First we rotate the xy-plane counterclockwise by 45

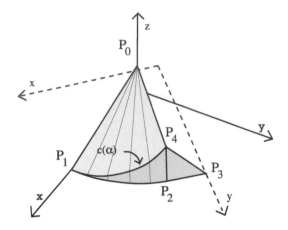

Figure 8. The fundamental region in a different coordinate system.

degrees and then translate it in the new x direction by $\sqrt{2}h$. This puts the point P_0 on the z-axis. Our new coordinates, x' and y' are derived from x and y by the following relations:

$$x' = \frac{x+y}{\sqrt{2}} - \sqrt{2}h \quad y' = \frac{-x+y}{\sqrt{2}}.$$

Finally, we convert from Cartesian coordinates (x', y', z) to cylindrical coordinates (r, θ, z) using the standard definitions:

$$r = \sqrt{(x')^2 + (y')^2} \ and \ \theta = \arctan{(y'/x')}.$$

Now we are ready to derive expressions for $r(\alpha)$, $\theta(\alpha)$, and $z(\alpha)$. Since our change of coordinates leaves z untouched, our earlier observation about $z(\alpha)$ remains valid:

$$z(\alpha) = 1 - \cos(\alpha).$$

The requirement that every point on c be at distance 1 from P_0 becomes:

$$r(\alpha)^2 + (z(\alpha) - h)^2 = 1$$

or

$$r(\alpha) = \sqrt{1 - (z(\alpha) - h)^2} = \sqrt{1 - (1 - \cos(\alpha) - h)^2}.$$

To find an expression for $\theta(\alpha)$, we recall from basic calculus the formula for an element of arc length α in cylindrical coordinates:

$$(dr)^2 + r^2(d\theta)^2 + (dz)^2 = (d\alpha)^2.$$

Combining this with our expressions for r and z, we get

$$\frac{d\theta}{d\alpha} = \frac{1}{r}\sqrt{1 - \left(\frac{dr}{d\alpha}\right)^2 - \left(\frac{dz}{d\alpha}\right)^2} = \frac{\sqrt{\cos^2(\alpha) - (1 - h - \cos(\alpha))^2}}{1 - (1 - h - \cos(\alpha))^2}.$$

Thus,

$$\theta(\alpha) = \int_0^\alpha \frac{\sqrt{\cos^2(\beta) - (1 - h - \cos(\beta))^2}}{1 - (1 - h - \cos(\beta))^2}\, d\beta.$$

This can be shown to be an elliptic integral and does not have a closed form solution.

Now we must show that there exists a value of h such that the endpoint of the curve at $c(\pi/4) = P_4$. The symmetry constraints described earlier are required to ensure that the points P_2 and P_4 will match up with corresponding points on neighboring fundamental regions. These constraints imply that

$$P_4 = \left(1 - \frac{1}{\sqrt{2}}, h + \sqrt{1 - 2\left(h - 1 + \frac{1}{\sqrt{2}}\right)^2}, 1 - \frac{1}{\sqrt{2}}\right)$$

in the original (x, y, z) coordinate system. Converting to cylindrical coordinates, we get

$$P_4(r) = \sqrt{1 - \left(1 - \frac{1}{\sqrt{2}} - h\right)^2} = r(\pi/4);$$

$$P_4(\theta) = \arctan\left(\frac{h - 1 + 1/\sqrt{2} + \sqrt{1 - 2(1 - h - 1/\sqrt{2})^2}}{h - 1 - 1/\sqrt{2} + \sqrt{1 - 2(1 - h - 1/\sqrt{2})^2}}\right);$$

$$P_4(z) = 1 - \frac{1}{\sqrt{2}} = z(\pi/4).$$

We see that $r(\pi/4) = P_4(r)$ and $z(\pi/4) = P_4(z)$, regardless of our choice of h. So we need to find h such that $\theta(\pi/4) = P_4(\theta)$.

I used Mathematica to evaluate these expressions numerically and found that a solution exists for $h \approx .2231$. For those who are not satisfied with numerical results, we could show that there exists h such that $\theta(\pi/4) - P_4(\theta)$ is positive, and another h for which it is negative, and invoke the Mean Value Theorem to prove the existence of a true solution. Hence, the Orb is valid.

There remains, however, one additional constraint that I have neglected to mention. It would be nice if the tangent vector at the endpoint of the curve $c'(\pi/4)$ lies in the plane of $\triangle P_2 P_3 P_4$. This would imply that the transition from the cylindrical section of the Orb to the planar section is smooth. It is not absolutely required for the validity of the Orb, however, since, if the tangent vector did not lie in the plane, we could "fix" the model by adding eight additional straight creases to each of the modules. Fortunately, the tangent vector does lie in the plane for the correct value of h, and the additional creases are not necessary. For the sake of brevity, I do not include the proof in this paper.

I intend, in the future, to investigate the validity of other origami models, and in particular, a larger set of Orbs constructed by making conical dimples at the vertices of other polyhedra besides the cube.

References

[1] Coxeter, H. S. M. *Introduction to Geometry*, Wiley and Sons, New York, 1969.

[2] Mitchell, David, *Mathematical Origami*, Tarquin Publications, Norfolk, England, 1997.

[3] Mosely, J., *Blanks for forming hollow objects etc.*, United States Patent No. 6,027,819, 2000.

Mathematical Origami: Another View of Alhazen's Optical Problem

Roger C. Alperin

1 Fields and Constructions

We can solve some elementary problems from geometry using origami foldings. Below are the axioms which guide the allowable constructible folds and points in $\mathbb{C}$, the field of complex numbers, starting from the labeled points 0 and 1 (see [1] for more details and references).

(1) The line connecting two constructible points can be folded.

(2) The point of coincidence of two fold lines is a constructible point.

(3) The perpendicular bisector of the segment connecting two constructible points can be folded.

(4) The line bisecting any given constructed angle can be folded.

(5) Given a fold line l and constructed points P, Q, then whenever possible, the line through Q, which reflects P onto l, can be folded.

(6) Given fold lines l, m and constructed points P, Q, then whenever possible, any line which simultaneously reflects P onto l and Q onto m, can be folded.

The first three axioms are the Thalian constructions which ensure that we have a field after we have a nonreal complex number z. Some properties of these constructions include the following: reflections can be folded; the set of slopes of fold lines correspond to the points constructed on the imaginary axis together with ∞; the points which can be constructed contain the field $\mathbb{Q}(z)$ and is contained in $\mathbb{Q}(z, i)$; and starting with $z = \frac{1+i\sqrt{3}}{2}$, the point i cannot be constructed.

Adding the fourth axiom gives the Pythagorean field of points constructible by straightedge and dividers as discussed by Hilbert. Some properties of this field include the following: the (real) field of Pythagorean constructible numbers is characterized as the smallest field containing the rational numbers and is closed under $\sqrt{a^2 + b^2}$ for (a, b) a constructed point; furthermore, any algebraic conjugate of a real Pythagorean is real and has degree which is a power of 2; the number $\sqrt{1 + \sqrt{2}}$ does not belong to the field; and any regular polygon which can be constructed by ruler and compass can be constructed with these axioms.

Adding the fifth axiom gives the Euclidean field of points constructible by straightedge and compass; it enables the construction of the tangents (through Q) to a parabola with focus P and directrix l. The Euclidean field is characteized as the smallest field containing the rational numbers and closed under all square roots.

Allowing axiom (6) yields the construction of the simultaneous tangents to two parabolas. Since this is related to pencils of conics we may also refer to the origami numbers as constructible by straightedge, compass and pencils. The axioms (1)–(6) are the origami construction axioms for the complex origami numbers, $\mathcal{O}$. The origami constructions (1)–(6) enable us to construct a real solution to a cubic equation with real coefficients in this field $\mathcal{O}$. To see this, consider the conics $(y - \frac{a}{2})^2 = 2bx$, $2y = x^2$. These conics have foci and directrices that are constructible using field operations involving a and b. Consider a simultaneous tangent, a line with slope μ meeting these curves at the respective points (x_0, y_0), (x_1, y_1). By differentiation, we find that the slope μ of a common tangent of these two parabolas satisfies $\mu^3 + a\mu + b = 0$, and hence we can solve any cubic equation with specified real constructible $a, b \in \mathcal{O}$ for its real roots. Using the resolvent cubic of a quartic equation, we can also solve fourth-degree equations over $\mathcal{O}$. The field of origami constructible numbers is characterized as the smallest subfield of the complex numbers which contains the rational numbers and is closed under all square roots, cube roots, and complex conjugation.

This origami field of numbers $\mathcal{O}$ is the same field Viete studied systematically at the beginning of "algebraic geometry" in 1600. Of course Viete was, in a sense, rediscovering what was already long ago known via *neusis* constructions, to Archimedes and Apollonius (250 BC), some of which was written down by Pappus more than five hundred years later (325 AD). One of Pappus' aims, it appears, was to prove that all conic constructibles are the same as *neusis* constructions. This was later picked up by Alhazen in the 11th century. Alhazen was interested in reconstruction of the lost works of Apollonius related to these geometrical constructions.

Moreover, the constructions using a *neusis* or marked ruler are the same constructions as those intersections of conics; both are equivalent to the origami constructions described by axioms (1)–(6). Our renewed interest in Alhazen's Problem arose because of the this equivalence [1].

2   Harmonic Origami Numbers

We can introduce yet another intermediate field of numbers and constructions using the Pythagorean axioms (1)–(4) together with

T. Given P and Q and a line l through P then we can simultaneously fold Q onto l and P onto the perpendicular bisector of PQ.

This gives the field of harmonic numbers. Some of the properties of this field include the following: the real subfield of harmonic constructible numbers is the smallest field closed under $\sqrt{a^2 + b^2}$ for a, b in the field and also closed under adjunction of any real number satisfying an irreducible cubic having three real roots with real harmonic number coefficients; trisections can be done using axiom T; and $\sqrt[3]{2}$ is not harmonic constructible.

Call an integer of the form $2^a 3^b$, a harmonic integer, so named by Phillip de Vitry (14th century) in studying relations to music. Any regular polygon with n sides can be constructed using these axioms iff $\Phi(n)$ is harmonic iff n is a product of a harmonic integer and distinct primes p, so that $p - 1$ is harmonic. Any regular polygon with n sides can be constructed iff $\Phi(n)$ is a harmonic integer; the real constructible harmonic points are characterized as the real origami numbers which have Galois closure whose degree is an harmonic integer. We shall leave the details of these remarks for a later time.

3   Some Elementary Problems from Geometry

To lead up to Alhazen's Problem, we start with an easy geometry problem, the river crossing.

Example 3.1: In going from town A to town B, one must cross the river L (of fixed width d) at a point z. Locate z so as to minimize the distance (by land) from A to B.

The solution can be folded easily; fold so that the the river disappears, down the middle and then fold the line AB; this line gives the path, and the bridge is the perpendicular at the river. It is clear that if we have any path from A to the river and across followed by the path to B that when we fold the river away that the two parts of the path give a length which can be shortened if it is made into a straight path.

A similar problem arises when we go from A to B on the same side of L.

Example 3.2: In going from town A to town B on the same side of the river, one must first stop at the river L at a point z. Locate z so as to minimize the distance from A to B.

One can first reflect B across the line L to B_0, then the straight line AB_0 meets the river at $z = C$, giving the stopping point for the shortest distance. Morever, now it is also apparent that the two angles at C made with L are equal.

After giving the solutions to these problems, I suggest to my students that it would be interesting to formulate and solve these problems for a circular rather than a straight river. I had in mind one point inside an inner circle and the other outside the outer concentric circle. The bridge is to be along the diameters.

Example 3.3: In going from town A to town B, one must cross the annular river (or moat) of fixed width d at a point z. Locate z so as to minimize the distance (by land) from A to B.

In this circular or annular problem after we locate the points $z = C$ on the outer circle and the corresponding point D on the inner circle which are the ends of the bridge, then we can fold away the section between them. What remains are two segments which give the path from A to B. In order to minimize this, we should make a straight path as in Figure 1. Thus we need only construct AB originally, and this gives the point C. Another potential solution occurs at the point where AB meets the inner circle.

The last problem, I have learned, is equivalent to the classical Alhazen's Problem. Suprisingly, several others have recently taken a look at this ancient piece of beautiful work. It is the circular analogue of the second problem. We formulate the problem as follows.

Example 3.4: Given points A, B exterior to a circle Γ, locate $z = C$ on Γ so that the distance $AC + CB$ is minimized.

At a point $z = C \in \Gamma$, we construct the tangent to Γ. Reflect B to the other side giving B_1. The sum of the two segment lengths AC and CB can be shortened by making $AC \cup CB_1$ straight. Thus the shortest distance occurs when $\angle ACB$ is bisected by the diameter through C. A similar angle restriction gives the shortest distance when the two points are first inside the

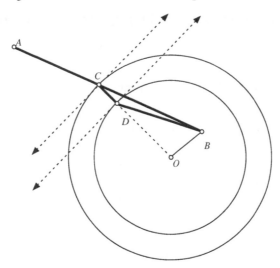

Figure 1. Bridge over annular river.

circle. (We leave other questions of this sort involving the other conics to the interested reader.)

When the points are outside the circle this is Alhazen's Problem, a problem of optics possibly first formulated by Ptolemy. The famous al-Hasen Ibn al-Haytham (latinized Alhazen) lived in the 11th century (Basra, Persia and Cairo, Egypt) and wrote a most influential text on optics, later used by Renaissance scientists. According to [7], Alhazen's Problem can be reduced to the construction by a *neusis*, ([7] Lemma 1, p. 310).

Huygens' solution (1672) of Alhazen's Problem is equivalent to Ibn al-Haytham's original solution [7]. Huygens was displeased with Alhazen's solution by a classical *neusis*. The problem was solved by Huygens, by reducing it to the intersection of the circle with an equilateral hyperbola; the solutions can then be realized as solutions of a real fourth-degree equation. A recent construction of this hyperbola is given in [5]. With the given points inside the circle, this is the circular billiard problem considered also in [8]. One can use inversion in the given circle to convert one formulation to the other. Yet another formulation [3], asks when the chords from a point on the circle to the given points are equal.

It is important to realize that geometrical constructions of the early Renaissance eventually led to the discovery of formulas for the roots of cubics and quartics involving radicals by del Ferro in the early 16th century and popularized by Cardano in the latter part of that century.

4   Alhazen's Problem–Huygens' Solution

We restate Alhazen's Problem in terms of angles.

Example 4.1: Given points A, B exterior to a circle Γ, locate $z = C$ on Γ so that $\angle ACB$ is bisected by the diameter through C.

We regard the given points $a = A$, $b = B$ as complex numbers and the circle Γ of radius 1 centered at the origin O. Using the *argument* of a complex number (or polar coordinates) as in [6, 8], $arg(\frac{a-z}{b-z})$ is the measure of $\angle azb$. Thus we want z so that $\angle azO = \angle Ozb$ or, equivalently, $arg(\frac{a-z}{O-z}) = arg(\frac{O-z}{b-z})$; we obtain a real number by evaluating $\frac{a-z}{z}(\frac{z}{b-z})^{-1}$. Using the fact that this number is equal to its conjugate and simplifying with $z\bar{z} = 1$, we find that

$$(ab)\bar{z}^2 - (\bar{a}\bar{b})z^2 = (a+b)\bar{z} - (a\mp b)z.$$

This gives an equation relating the imaginary parts of two complex numbers,

$$\mathcal{I}m((ab)\bar{z}^2) = \mathcal{I}m((a+b)\bar{z}).$$

This is easily seen to be satisfied by

$$z \in \{O, \frac{1}{\bar{a}}, \frac{1}{\bar{b}}, \frac{1}{\bar{a}} + \frac{1}{\bar{b}}\}.$$

Let $p = Im(ab)$, $q = Re(ab)$, $r = Re(a+b)$, $s = Im(a+b)$. We can write this equation in terms of real coordinates using $z = x + y \cdot i$ to obtain

$$p(x^2 - y^2) - 2qxy = sx - ry.$$

This is the equation of an hyperbola. Thus, the solution to Alhazen's Problem, if it exists, occurs at one of the intersections of this hyperbola and the unit circle centered at the origin.

Now rotate the plane about the origin by the angle so that the positive x-axis bisects $\angle a0b$, hence $b = k\bar{a}$ for some $k > 0$; consequently $p = Im(ab) = 0$, thus eliminating the terms x^2, y^2 from the equation of the hyperbola above. The equation of the equilateral hyperbola now simplifies to

$$(x - \frac{r}{2q})(y + \frac{s}{2q}) = -\frac{rs}{4q^2}.$$

It follows that the center of this hyperbola is at

$$(\frac{r}{2q}, \frac{-s}{2q}) = (\frac{Re(a+b)}{2ab}, \frac{-Im(a+b)}{2ab}) = \frac{1}{2}(\frac{1}{\bar{a}} + \frac{1}{\bar{b}}).$$

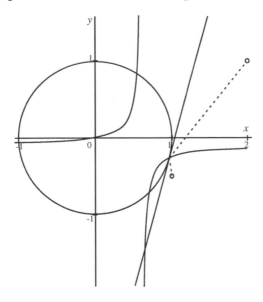

Figure 2. Huygens' solution: $A = (2, 1), B = (1, -\frac{1}{2})$.

A simple example for $a = 2 + i, k = \frac{1}{2}$ is displayed in Figure 2. The point $z \approx .9643353545 - .2646834413 \cdot i$ gives the solution to Alhazen's Problem where the x-coordinate solves $x^2 + (\frac{-x}{2(5x-3)})^2 - 1$ and the y-coordinate is $-\sqrt{1 - x^2}$. The $arg(\frac{a-z}{b-z})$ gives 132.06736 degrees, which is twice $arg(\frac{a-z}{z})$.

The solution to Alhazen's Problem is the same as the simultaneous solution to

$$A_1 : \; x^2 + y^2 = 1, \quad A_2 : \; 2qxy + sx - ry = 0.$$

These simultaneous solutions to A_1 and A_2 also lie on any curve in the pencil $A_2 - \lambda A_1$; the pencil has matrix

$$\begin{pmatrix} -l & q & \frac{s}{2} \\ q & -l & -\frac{r}{2} \\ \frac{s}{2} & -\frac{r}{2} & l \end{pmatrix}.$$

The determinant of this matrix pencil gives a cubic polynomial, $p(l) = l^3 + \frac{1}{4}(s^2 + r^2 - 4q^2)l - \frac{1}{2}rqs$. Now solve the reduced cubic $p(l) = 0$ using the technique allowed by axiom (6). Finding the roots to this cubic is similar to the technique of solving the resolvant cubic of a quartic. Each root gives a degenerate conic in the pencil: a pair of lines meeting at a diagonal point. One needs to use square roots to obtain the equations of these lines. By solving this cubic and the quadratics, we obtain the (at most) six lines in the complete

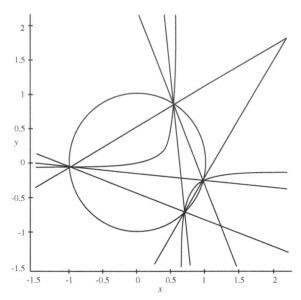

Figure 3. Complete quadrilateral.

quadrilateral, and therefore, the four points of intersection of the conics and the three diagonal points as in Figure 3 for the example above.

We can also replace the Alhazen-Huygens solution as simultaneous intersection of conics by a single real quartic. We describe in the next section how pencils and origami may be used to simplify and solve the quartic.

5   An Origamic Solution to the Quartic

We assume that we are given a general real quartic. Complete the quartic to eliminate the coefficient of x^3 to obtain say, $x^4 - 2cx^2 - dx - e = 0$ for certain real c, d, e. (If $d = 0$, then we can solve the quartic using square roots so we assume $d \neq 0$.) Put $y = x^2 - c$; then the quartics' solutions are the same as the simultaneous solutions to two parabolas

$$P_1 : \ y = x^2 - c, \ P_2 : \ y^2 = dx + e + c^2.$$

The duals of these two parabolas give two conics. The dual equations are

$$C_1 : \ X^2 + 4cY^2 - 4Y = 0, \ C_2 : \ 4(e + c^2)X^2 + d^2Y^2 - 4dX = 0.$$

Duality Principles

1. The common tangents to C_1 and C_2 correspond to the common points of P_1 and P_2.

2. The common tangents to P_1 and P_2 correspond to the commmon points of C_1 and C_2.

When a tangent line's equation is, say, $aX + bY + 1 = 0$, then a corresponding common point on the dual is (a, b) and conversely.

Now we consider the pencil generated by C_1 and C_2; it is $C_2 - uC_1$; when $u = 4(e + c^2)$ we obtain the parabola C_3. Next consider $C_3 - vC_2$; when $v = \frac{d^2 - 4cu}{d^2}$ we obtain an independant parabola C_4.

In this case, we have

$$C_3 : \ d^2 v Y^2 + 4uY = 4dX, \ \ C_4 : \ dvX^2 + 16cX = 4dY,$$

and their duals

$$P_3 : \ (ux + dy)^2 = d^3 vx, \ \ P_4 : \ (dx + 4cy)^2 = d^2 vy,$$

are parabolas since the equations have double points at infinity.

By folding tangents to C_3 and C_4 using axiom (6), we can recover the points of intersection on P_3 and P_4 by duality. We obtain the common points of P_1 and P_2 by use of the associated complete quadrilateral which has the same diagonal points as the pencil P_3 and P_4. We shall elaborate on this connection in further detail at another time.

6   Geometry and Axiomatic Origami

One interesting problem is to give elegant constructions using a given set of axioms. For example, show how to solve problems from Euclidean geometry with the Euclidean axioms or Pythagorean axioms. With the origami axioms, we can fold the common intersections or tangents into two conics. Is there an elegant solution?

Example 6.1: Fold the intersections of a conic and a line. Fold the tangents to a conic from given point. Fold the common tangents to two conics. Fold the common points of two conics.

One needs to give geometrical data about a conic. Specifying a conic requires five pieces of information, either points on the curve or tangent lines to the curve; this can also be replaced by focus or directrix information. For example, the circle is given either by three points (and two circular points at infinity) or three tangents, or the center and tangents. A parabola can be given by four points (and a tangent line at infinity), or a focus and directrix; hyperbolas or ellipses can be specified by data from foci, center, diameter, tangents, points, or asymptotes.

Example 6.2: Given a square sheet of paper, fold a regular n-gon of largest possible area using one of the Pythagorean, harmonic, Euclidean, or origami axiom systems?

To fold a heptagon using harmonic constructions, take a piece of paper, say, twelve units square. Fold the corner A to a point on CV using a fold through the corner F as shown in the diagram in Figure 4. This fold meets the central vertical line at a point; the midpoint B_1 of this point and the top edge is constructed. The distance of B_1 from V is $3\sqrt{3}$, and so the hypotenuse

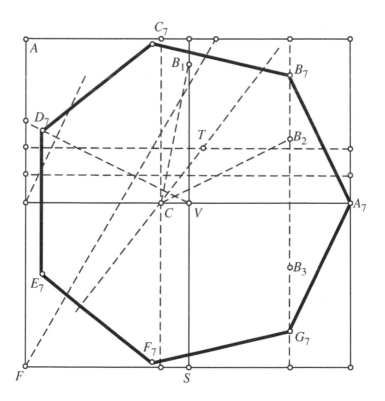

Figure 4. Folding the regular heptagon using harmonics.

with the edge CV of length 1 has length $\sqrt{28}$. Now trisect $\angle B_1CV$ to get the point T as in Abe's trisection (it is the trisection with the base CB_1). Reflect B_1 to B_2 across CT. Reflect B_2 across CV to get B_3. Now fold A_7 onto the line B_2B_3 passing through V. This gives the second point of the heptagon B_7. Now reflect this across VA_7 to get the vertex G_7 of the heptagon. Fold B_7A_7 across the line VB_1, and bisect the angle with CV; construct the vertex D_7 on this bisector. Reflect across CV to get vertex E_7. Bisect $\angle E_7VG_7$; construct F_7 on that bisector. Reflect across CV to get the final vertex.

As one can show, if a polygon can be folded in the square, then so can the optimal one. An analysis of the cosine of the angle involved in tipping the polygon to its optimal position shows that it belongs to the field [2]. Can *you* fold the optimal heptagon using only harmonic constructions?

References

[1] Roger C. Alperin, *A Mathematical Theory of Origami Constructions and Numbers*, New York J. Math., 6, 119-133 (2000), http://nyjm.albany.edu:8000/j/2000/6-8.html.

[2] David Dureisseix, *Searching for the Optimal Polygon*, webprint (October 1997).

[3] Jack M. Elkin, *A Deceptively Easy Problem*, Math. Teacher, 194-199 (March 1965).

[4] W. Gander and D. Gruntz, *The Billiard Problem*, Maple Worksheet Shareware (July 1992) gander@inf.ethz.ch.

[5] Norbert Hungerbühler, *Geometrical Aspects of the Circular Billiard*, El. Math. 47, 114-117 (1992).

[6] Peter M. Neumann, *Reflections on Reflection in a Spherical Mirror*, Am. Math. Monthly, 105, No. 6, 523-528 (1998).

[7] A. I. Sabra, *Ibn al-Haytham's Lemmas for Solving "Alhazen's Problem"*, Archive for History of Exact Sciences, 26, 299-324 (1982).

[8] Jörg Waldvogel, *The Problem of the Circular Billiard*, El. Math. 47, 108-113 (1992).

Just Like Young Gauss Playing with a Square: Folding the Regular 17-gon

Robert Geretschläger

When we compare Euclidean constructions to constructions applying other sets of rules, one point of particular interest is how the rules measure up when applied to the various construction problems that have been on mathematicians' minds since antiquity. Some of these problems, such as the squaring of the circle or the trisection of general angles, are now known to be impossible to solve using Euclidean methods, but they may be solved by applying other rules of construction. Others, such as determining tangent lines of conics, are known to be quite straightforward when applying Euclidean methods, and may or may not be accessible by other methods, depending on the definition of the rules.

As was shown in [6], we know that a reasonable definition of origami constructions allows any construction problem that can be solved by Euclidean methods to be solved by origami methods. In fact, each Euclidean step can be replaced by a single origami step or a specific combination of origami steps. Of the classic construction problems of Greek antiquity, the one that stands out from the rest when considering origami constructions is the problem of constructing regular polygons. One reason for this is the practical applicability of some simple regular polygons when producing folding bases or modules, but another is certainly the fact that most origami starts from a square, and therefore from one of the simplest regular polygons. Folding a different regular polygon from this starting point becomes interesting for more abstract reasons than mere practicality.

Methods for folding regular polygons have appeared in the literature numerous times. Since they are frequently meant to be used in practical applications, theoretical precision is often sacrificed in favor of simplicity, and so many of these methods are only approximations when viewed from a more mathematical vantage point. There have, however, been numerous examples of theoretically exact folding methods for regular polygons developed, some nice examples of which are described in [3], [4], [5], [7], [8], [9], [10], [12], [13] and [16]. The earliest such methods in the modern literature are almost certainly to be found in T. Sundara Row's classic [15], although it can be argued that Euclid's methods for constructing polygons (the equilateral triangle, square, regular pentagon, and regular hexagon), which predate Row by several millenia, can readily be "translated" to origami constructions.

Folding a regular 17-gon is of particular interest, since this is an n-gon with prime n that can be constructed with straight-edge and compass, as was shown so brilliantly by the young Gauss in the late 18th century. Unlike the 7-gon or the 11-gon, for example, the equations associated with the 17-gon can all be reduced to linear and quadratic problems, making them accessible to Euclidean constructions.

Gauss' result was merely a by-product of a much more elegant theorem, which implied that any regular n-gon can be constructed by Euclidean methods if n is a so-called *Fermat* prime, i.e., a prime of the form $n = 2^{2^k} + 1$. Since 17 is indeed prime, and $17 = 2^{2^2} + 1$, the constructability of the 17-gon follows. This result, along with many others from Gauss' dissertation was later published in his *Disquisitiones Arithmeticae*. Until this time, there had been no reason to assume that the construction of the heptadecagon was possible. Constructions for the regular n-gon were well-known from antiquity for $n = 3, 4, 5$, and 6, but a construction had eluded the Greeks for the seemingly simpler case of $n = 7$, and so there was no reason to think that there should be one for $n = 17$. Young Gauss' discovery changed all that.

Many different Euclidean constructions of the 17-gon have since been proposed. Perhaps the most elegant is the following, due to H.W. Richmond [14].

We choose a point O as midpoint of the circumcircle of the 17-gon, and a point 1 as one vertex. The circumcircle then passes through 1. We construct the diameter of the circumcircle through 1 and a radius OA perpendicular to $O1$. Point B is then constructed on OA such that the line segment OB is one quarter the length of OA. Drawing the line segment $B1$, we next determine the point C on $O1$ such that $\angle OBC = \frac{1}{4}\angle OB1$, and then the point D on the diameter of the circumcircle through 1 such that $\angle CBD = \frac{\pi}{4}$. The circle with diameter $D1$ intersects OA in a point E, and intersecting the circle with midpoint C and passing though E with the diameter of the circumcircle through

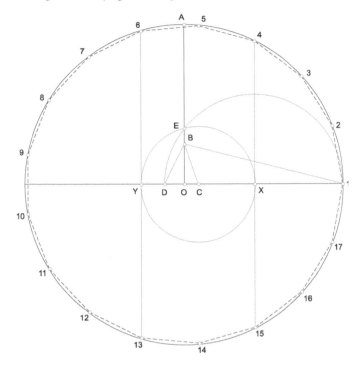

Figure 1. Constructing the 17-gon.

1 yields points X and Y. The lines perpendicular to the diameter and passing through X and Y intersect the circumcircle in vertices 4 and 15, and 6 and 13, respectively. The 17-gon can then easily be completed by applying the various symmetries inherent in the regular 17-gon.

Since we can replace any Euclidean steps in such a construction by combinations of origami steps, it must be possible to modify this construction of the regular 17-gon to one using origami methods. This results in a somewhat unwieldy folding procedure, however. To develop a more practicable method, an idea similar to that applied in [7], [8], [9] and [10] for various other polygons seems more appropriate.

We think of the vertices of the 17-gon as the solutions of a polynomial equation in the complex plane. In this case, the equation in question is

$$z^{17} - 1 = 0.$$

We note that the unit circle is the circumcircle of the 17-gon, and that the point $z_1 = 1$ on the real axis is one vertex. The other 16 vertices are therefore the roots of the equation

$$\frac{z^{17} - 1}{z - 1} = z^{16} + z^{15} + \ldots + z + 1 = 0,$$

and we can write all 17 roots (in polar coordinates) as

$$z_k = \left(1; \frac{2\pi(k-1)}{17}\right) = \cos\frac{2\pi(k-1)}{17} + i \cdot \sin\frac{2\pi(k-1)}{17}$$

for $k = 1, 2, \ldots, 17$.

Naming $z_2 =: \zeta$, we then have $z_k = \zeta^{k-1}$ for $k = 2, 3, \ldots, 17$. Grouping the powers of ζ in an appropriate way allows us to reduce the solution of the degree 16 equation to the successive solution of a number of equations of lesser degree. Indeed, as we know from Gauss, if we do it correctly, we can expect all of these equations to be of no greater degree than two.

Successive squaring of ζ yields values of z_k in the order $\zeta, \zeta^2, \zeta^4, \zeta^8, \zeta^{16}, \zeta^{15}, \zeta^{13}, \zeta^9$, and the powers of ζ missing in this sequence are obtained by successive squaring of ζ^3 in the order $\zeta^3, \zeta^6, \zeta^{12}, \zeta^7, \zeta^{14}, \zeta^{11}, \zeta^5, \zeta^{10}$. We define variables y_1 and y_2 as

$$y_1 = \zeta + \zeta^2 + \zeta^4 + \zeta^8 + \zeta^{16} + \zeta^{15} + \zeta^{13} + \zeta^9$$

and

$$y_2 = \zeta^3 + \zeta^6 + \zeta^{12} + \zeta^7 + \zeta^{14} + \zeta^{11} + \zeta^5 + \zeta^{10},$$

and note that we then have

$$y_1 + y_2 = \zeta + \zeta^2 + \zeta^3 + \ldots + \zeta^{16} = -1$$

and

$$y_1 \cdot y_2 = 4 \cdot (\zeta + \zeta^2 + \zeta^3 + \ldots + \zeta^{16}) = -4.$$

This means that y_1 and y_2 are the solutions of the equation $y^2 + y - 4 = 0$.

Of these two solutions, y_1 is certainly positive (as we see in Figure 2), since $\zeta + \zeta^{16}$, $\zeta^2 + \zeta^{15}$ and $\zeta^4 + \zeta^{13}$ are all positive real numbers and only $\zeta^8 + \zeta^9$ is negative. We therefore see that

$$y_1 = \frac{-1 + \sqrt{17}}{2} \qquad \text{and} \qquad y_2 = \frac{-1 - \sqrt{17}}{2}.$$

In the next step, we define variables n_1 and n_2 by $n_1 = \zeta + \zeta^4 + \zeta^{16} + \zeta^{13}$ and $n_2 = \zeta^2 + \zeta^8 + \zeta^{15} + \zeta^9$. We have

$$n_1 + n_2 = \zeta + \zeta^2 + \zeta^4 + \zeta^8 + \zeta^{16} + \zeta^{15} + \zeta^{13} + \zeta^9 = y_1$$

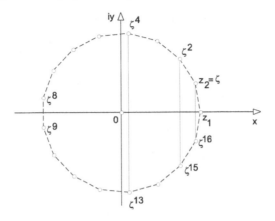

Figure 2. Roots of $z^{17} - 1$ in the complex plane.

and

$$n_1 \cdot n_2 = \zeta + \zeta^2 + \zeta^3 + \ldots + \zeta^{16} = -1.$$

These two numbers are therefore the solutions of the quadratic equation $n^2 - y_1 n - 1 = 0$, and glancing at Figure 2 shows us immediately that n_1 is positive (since both $\zeta + \zeta^{16}$ and $\zeta^4 + \zeta^{13}$ are positive) and n_2 is negative (since $\zeta^8 + \zeta^9$ is negative and much nearer to $2 \cdot (-1)$ than $\zeta^2 + \zeta^{15}$ is to $2 \cdot (+1)$).

Furthermore, we define m_1 and m_2 in an analogous way by $m_1 = \zeta^3 + \zeta^{12} + \zeta^{14} + \zeta^5$ and $m_2 = \zeta^6 + \zeta^7 + \zeta^{11} + \zeta^{10}$. We have

$$m_1 + m_2 = \zeta^3 + \zeta^6 + \zeta^{12} + \zeta^7 + \zeta^{14} + \zeta^{11} + \zeta^5 + \zeta^{10} = y_2$$

and

$$m_1 \cdot m_2 = \zeta + \zeta^2 + \zeta^3 + \ldots + \zeta^{16} = -1,$$

and these numbers are the solutions of the quadratic equation $m^2 - y_2 m - 1 = 0$. It is obvious that m_2 is negative, since both $\zeta^6 + \zeta^{11}$ and $\zeta^7 + \zeta^{10}$ are negative real numbers. The other solution m_1 is positive, since $\zeta^3 + \zeta^{14}$ is positive with $\mathrm{Re}\, \zeta^3 > |\mathrm{Re}\, \zeta^5|$.

As a final step, we define variables v_1 and v_2 as $v_1 = \zeta^2 + \zeta^{15}$ and $v_2 = \zeta^8 + \zeta^9$. Since we have

$$v_1 + v_2 = \zeta^2 + \zeta^8 + \zeta^{15} + \zeta^9 = n_2$$

and

$$v_1 \cdot v_2 = \zeta^6 + \zeta^7 + \zeta^{11} + \zeta^{10} = m_2,$$

these are the solutions of the quadratic equation $v^2 - n_2 v + m_2 = 0$, with

$$v_1 = 2 \cdot \cos \frac{4\pi}{17} > 0 \text{ and } v_2 = 2 \cdot \cos \frac{16\pi}{17} < 0.$$

In the following folding procedure, we shall therefore strive to determine v_1 and then go on to complete the 17-gon by applying the various symmetries inherent to it.

1   The Regular 17-gon

1

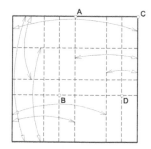

Fold edge-to-edge and unfold twice; fold edges to creases and unfold to create six more creases.

2

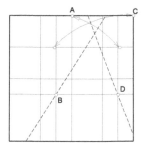

Fold A to horizontal crease such that resulting crease passes through B and unfold; do same with C such that resulting crease passes through D.

3

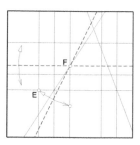

Fold horizontal crease through F and unfold; then fold E onto vertical crease through F such that new crease also passes through F and unfold.

4

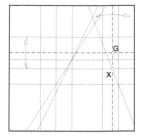

Fold vertical crease through X and unfold; fold one horizontal crease onto the other and unfold; these two resulting creases intersect at point G.

5

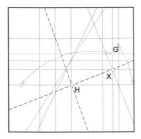

Fold crease from Step 2 through D onto itself through X and unfold; resulting crease intersects horizontal crease through E at point H; fold G onto EH such that crease passes through H and unfold.

6

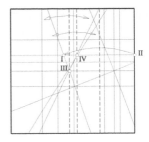

Fold vertical creases through III and IV and unfold; fold II onto I.

7

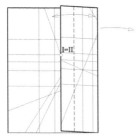

Fold both layers through crease from Step 1 through D and unfold everything.

8

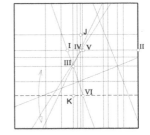

Fold horizontal crease through K and unfold.

9

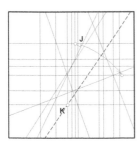

Fold J onto first horizontal crease from Step 1 such that crease passes through K and unfold.

10

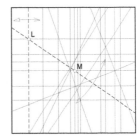

Fold crease from Step 9 onto itself through M and unfold; fold vertical crease through L and unfold.

11
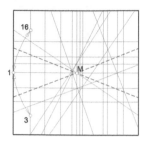

Fold point 1 onto crease from previous step and unfold twice.

12
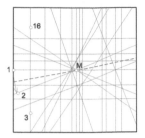

Fold 1 onto crease from previous step.

13
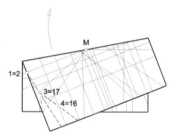

Fold back sides 23, 34, 171 and 1617 of 17-gon using points in both levels as guide; unfold first fold.

14
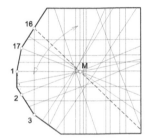

Fold back side 12; fold crease joining M and vertex 16; fold back sides of 17-gon using sides 1617, 171, 12, 23 and 34 as guidelines; unfold first fold.

15
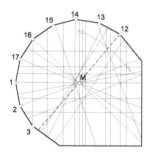

Fold crease joining M and vertex 12; fold back remaining sides of 17-gon using sides as guidelines; unfold first fold.

16
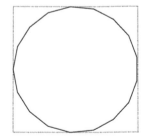

The regular 17-gon.

Why does this procedure work? In order to understand this, we first note that the solutions of any quadratic equation $x^2 + px + q = 0$ are the slopes of the tangents of the parabola with focus $F(0,1)$ and directrix $\ell : y = -1$ passing through the point $P(-p, q)$, which we can determine by folding F onto ℓ such that the crease passes through P. This method is presented in [11], and is derived in the following way.

Consider a parabola with the equation

$$p_o : \quad x^2 = 2u \cdot y$$

(where u is the parameter of p_o), and the point $P_o(v, w)$, as in Figure 3.

The line containing P_o and with slope s is described by the equation $y = s(x - v) + w$. If such a line is to be a tangent of the parabola p_o, the equation

$$x^2 = 2u(s(x - v) + w) \quad \Leftrightarrow \quad x^2 - 2usx + 2uvs - 2uw = 0$$

must have a discriminant equal to 0. This means

$$u^2 s^2 - 2uvs + 2uw = 0 \quad \Leftrightarrow \quad s^2 - \frac{2v}{u} \cdot s + \frac{2w}{u} = 0,$$

and we see that the slopes of the tangents of p_o containing P_o are the solutions of the equation $x^2 + px + q = 0$, if $p = -\frac{2v}{u}$ and $q = \frac{2w}{u}$ hold. If we therefore wish to determine the solutions of a given quadratic equation $x^2 + px + q = 0$, we can choose $u = 2$, $v = -p$, and $w = q$, and thus obtain the solutions as the slopes of the tangents of the parabola $p_o : x^2 = 4y$ containing the point $P_o(-p, q)$.

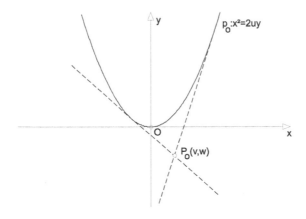

Figure 3. Finding tangents to a parabola.

These tangents are simply the creases obtained by folding the focus $F(0,1)$ onto directrix $\ell : y = -1$, such that the fold contains $P_o(-p, q)$, as stated.

We are now ready to see where this idea has been used in the folding procedure. In order to determine v_1 as the positive solution of the equation $v^2 - n_2 v + m_2 = 0$, we must first determine the values of n_2 and m_2, so that we can find the point with coordinates (n_2, m_2). We know that n_2 is the negative solution of $n^2 - y_1 n - 1 = 0$, and that m_2 is the negative solution of $m^2 - y_2 m - 1 = 0$, and as a first step, it is therefore necessary to determine y_1 and y_2 as the solutions of $y^2 + y - 4 = 0$. This is done in Steps 1 and 2. We assume that the folding square has sides of length 8 units, and due to the successive bisections in Step 1, we can assume that A has the coordinates $(0,1)$ and B has the coordinates $(-1, -4)$, whereby we assume that the x-axis is oriented to the right and the y-axis upwards. The line parallel to the upper edge and just below it is then represented by the equation $y = -1$, and folding A onto this line such that the crease passes through B in Step 2 yields a crease with slope y_1.

Since the size of the folding square is limited, we cannot fold a crease with slope y_2 in this way, but since points C and D have the same relative position as A and B, we can also assume that C has the coordinates $(0,1)$ and D has the coordinates $(-1, -4)$, and folding C onto the horizontal line $y = -1$ such that the crease passes through D then yields a crease with slope y_2.

In Step 3, we fold a crease with slope n_2. Since n_2 solves the equation $n^2 - y_1 n - 1 = 0$, we must determine points $F(0,1)$ and $P(y_1, -1)$ relative to some system of coordinates. This is done by assuming the x-axis of such a system is oriented upwards and the y-axis is oriented to the left. Leaving the unit length as in Steps 1 and 2, the choice of E in the grid point from Step 1 as shown yields a point F whose x-coordinate is larger than that of E by 2 units. We can therefore assume that the coordinates of E relative to the new system are $(0,1)$, and the coordinates of F in this system are therefore $(y_1, -1)$ as required, since F is one unit to the right of B, and therefore y_1 units higher than EB. Folding E onto the vertical crease through F (the line represented by the equation $y = -1$ in this system of coordinates) then yields a crease with slope n_2 in this system.

In Steps 4 and 5, we fold a crease with slope m_2, retaining the orientation and unit length of the system of coordinates as in Steps 1 and 2. By first folding the crease with slope y_2 onto itself, we obtain a crease with slope $-\frac{1}{y_2}$. The x-coordinate of the point H is therefore smaller than that of G by $|y_2|$ (recall that y_2 is negative), and since G is located on the horizontal crease two units above that through H, we can assume that G has the coordinates $(0,1)$ and H has the coordinates $(y_2, 1)$. We know that m_2 is the negative solution of the quadratic equation $m^2 - y_2 m - 1 = 0$, and folding G onto the horizontal crease through H (which is represented by the equation $y = -1$) such that the crease passes through H therefore yields a crease with slope m_2.

We are now ready to determine v_1. Since v_1 is a solution of the quadratic equation $v^2 - n_2 v + m_2 = 0$, it appears as the slope of a crease derived by folding $(0, 1)$ onto the line represented by $y = -1$ such that the crease passes through the point (n_2, m_2) in some system of coordinates. This is done in Steps 6 through 9.

The horizontal crease three units from the upper edge intersects the crease with slope m_2 in I and the right edge of the folding square in II. Folding II to I in Step 6 and then refolding the crease one unit from the edge of the square through both layers in Step 7 yields a vertical crease intersecting III in V, and the crease with slope m_2 in VI. Since I and V are one unit apart, V and VI must be $|m_2|$ units apart, and the horizontal crease in Step 8 through VI is m_2 units below III.

At the same time, III intersects the crease with slope n_2 (relative to the rotated system of coordinates) in IV. The horizontal line one unit below III intersects this crease in III, and the vertical creases through III and IV are therefore $|n_2|$ units apart. It follows that the point K, in which the horizontal line through VI and the vertical line through III intersect, is $|n_2|$ to the left and $|m_2|$ below the point IV. Taking IV as the origin of a new system of coordinates, once again with the x-axis oriented to the right and the y-axis oriented upwards, K has the coordinates (n_2, m_2). The point J, in which the vertical line through IV intersects the horizontal line one unit above III, has the coordinates $(0, 1)$, and the horizontal line one unit below III is represented by the equation $y = -1$. The crease obtained by folding J onto this line such that the crease passes through K therefore solves the equation, and its slope is

$$v_1 = 2 \cos \frac{4\pi}{17}.$$

The crease perpendicular to this one through the midpoint M of the folding square in Step 10 therefore has the slope $-\frac{1}{2\cos\frac{4\pi}{17}}$. Since we wish the circumcircle of the 17-gon to be the in-circle of the folding square, we now assume that the edges of the square are two units in length. The distance between the midpoint M of the folding square and the horizontal line through J is then one half unit, and this line intersects the crease with slope $-\frac{1}{2\cos\frac{4\pi}{17}}$ in a point L, whose distance from the vertical line through the midpoint of the square is $\cos \frac{4\pi}{17}$.

If we assume in Step 11 that the point 1 in which the horizontal line through the midpoint of the square intersects the lefthand edge is a vertex of the 17-gon, we obtain vertices 3 and 16 by folding 1 onto the vertical line through L. The creases resulting from these folds are $M2$ and $M17$. Steps 12 through 15 then complete the 17-gon by applying its symmetries.

Acknowledgements: Many thanks are due to Michael Hofer and Stephen Keeling for their numerous helpful comments on the first version of this paper.

References

[1] W.W. Rouse Ball, H.S.M. Coxeter, *Mathematical Recreations and Essays*, 13th edition, Dover Publications Inc., New York, 1987.

[2] Benjamin Bold, *Famous Problems of Geometry and How to Solve Them*, Dover Publications Inc., Mineola NY, 1969.

[3] David Dureisseix, *Searching for optimal polygon*, unpublished manuscript, 1997.

[4] David Dureisseix, *Searching for optimal polygon - application to the pentagon case*, unpublished manuscript, 1997.

[5] David Dureisseix, *Searching for optimal polygon - remarks about a general construction and application to heptagon and nonagon*, unpublished manuscript, 1997.

[6] Robert Geretschläger, *Euclidean Constructions and the Geometry of Origami*, Mathematics Magazine, Vol. 68, No. 5, December 1995, pp. 357-371.

[7] Robert Geretschläger, *Folding the Regular Heptagon*, Crux Mathematicorum with Mathematical Mayhem, Vol. 23, Nr. 2, March 1997, pp. 81-88.

[8] Robert Geretschläger, *Folding the Regular Nonagon*, Crux Mathematicorum with Mathematical Mayhem, Vol. 23, Nr. 4, May 1997, pp. 210-217.

[9] Robert Geretschläger, *Folding the Regular Triskaidekagon*, unpublished manuscript, 1998.

[10] Robert Geretschläger, *Folding the Regular 19-gon*, unpublished manuscript, 1998.

[11] Robert Geretschläger, *Geometric Constructions in Origami*, to appear.

[12] Miyuki Kawamura, *Polyhedra by Origami*, Japan, 1995.

[13] Roberto Morassi, *The Elusive Pentagon*, Proceedings of the First International Meeting of Origami Science and Technology, Ferarra, 1989.

[14] H.W. Richmond, *A Construction for a Regular Polygon of Seventeen Sides*, Quart. J. Pure Appl. Math. 26, 1893, pp. 206-207.

[15] T. Sundara Row, *Geometric Exercises in Paper Folding*, Dover Publications Inc., New York, 1966, reprint of 1905 edition.

[16] Benedetto Scimemi, *Draw of a Regular Heptagon by the Folding*, Proceedings of the First International Meeting of Origami Science and Technology, Ferarra, 1989.

[17] Dirk J. Struik, *A Concise History of Mathematics*, Dover Publications Inc., New York, 1987.

Paper-Folding Constructions in Euclidean Geometry: An Exercise in Thrift

Benedetto Scimemi

1 Generalities

Let Σ denote an initial set of points and lines of the Euclidean plane.

A *paper-folding (PF) construction from* Σ is a finite chain of points and lines $E_0, E_1, ..., E_{n-1}, E_n$ where each E_i is either initial (belongs to Σ) or new (obtained from pre-existing objects $E_j, ..., E_k$ for some $j, ..., k < i$ by applying one of the following permitted procedures):

A new point is either

(1) the intersection of two pre-existing (PE) lines or

(2) the point where a PE point is carried by reflection (i.e., folding) on a PE line (i.e., crease).

A new line is either

(3) the line through two PE points or

(4) a line which simultaneously reflects two PE points onto (some points of) two PE lines, respectively.

A point or line E is *paper-folding constructible from* Σ if $E = E_n$ for some PF construction from Σ.

Paper-folding constructions have been introduced as a method—of both theoretical and practical relevance—to replace and "surpass" ruler and compass in solving classical problems, such as trisecting an angle, duplicating a cube, etc. In fact, one easily sees that the following constructions can be obtained as special cases of (4):

(5) the perpendicular bisector of a PE segment AB

(6) the angle bisectors of two PE lines

(7) the line through a PE point, perpendicular to a PE line

(8) the line through a PE point, which reflects another PE point onto some point of a PE line.

This suffices to show that all ruler-and-compass constructions can be accomplished by paper-folding. However, it can also be proved that procedure (4), if properly applied, actually solves all problems of algebraic degree 3. More precisely, the following characterization is well-known (see [1]): If K_0 denotes the field of the coordinates and coefficients of the initial objects (line equations should be normalized), then a point or line E is PF constructible if and only if there exists a finite chain of fields $K_0 < K_1 < ... < K_{m-1} < K_m$ such that the final field K_m contains the coordinates or coefficients of E and all the intermediate extensions have degree $[K_{i+1} : K_i] \leq 3$.

Having established the range of possible constructions, we want to take a different standpoint in this paper, which has to do with the economy of a solution rather than its abstract existence: How many steps does a constructor need in order to achieve a specific object? This question becomes relevant if we consider producing real creases in a sheet of paper, since each fold obviously introduces some error, and a long sequence of folding will drive one far from any acceptable precision. It also meets the mathematician's taste for sparing superfluous steps, thus gaining in neatness and elegance. One may therefore introduce the weight of a construction, not a precise mathematical concept, but an estimate of how long (or heavy) a certain construction is. We can tentatively measure this weight by the number of steps required. All procedures (1), (2), ..., (8) can be assumed to be constructions of weight one, by definition. Constructions of higher weight are illustrated by the following examples:

Example 1: The line through a given point, parallel to a given line. This construction has weight 2, since we can obtain it by applying (7) twice.

Example 2: The point symmetric to a given point with respect to another given point. We can apply the sequence of procedures (1), (7), (2): weight 3.

Example 3: An equilateral triangle on a given side: Apply (5), (8), and (2).

Example 4: The trisection of an angle. This is Abe's original idea (see [1]), which gave birth to the whole subject of PF constructions. The crucial step is obviously (4), but one needs three or four previous steps (depending on the way the angle is assigned) in order to provide proper pairs of points and lines to which (4) is applied. Moreover, the trisecting lines are perpendicular to the crucial creases, so that, altogether, this construction has weight about 5–6.

Example 5: A regular heptagon is constructed in [1] by calculating the minimal polynomial of $\cos(2\pi/7)$ and representing its roots graphically according to a general method which we owe to Lill. That construction implies at least ten steps before (4) can be applied, then four to five further steps to construct the final polygon vertices. The next section will propose an alternative construction of much smaller weight.

In what follows, we shall solve a number of classical problems of elementary geometry by paper-folding constructions. The suggested constructions claim to be relatively light, although we shall soon give up counting their weights. With the exception of the first and the last problems, all the remaining constructions regard extremal points on a conic curve, namely points P on the curve which produce either a (locally) maximal or minimal distance $|PK|$ from a given point K. We underline the fact that the curve will never be used as such, but only be identified by some of its characteristic points and lines (center, focus, directrix, asymptotes).

In order to prove that the suggested constructions actually work, rather than trying to reduce one problem to the other or to adopt a single technique, we have chosen to use a variety of elementary methods often based on analytic geometry. The reader is invited to find, if possible, lighter constructions and alternative proofs entirely based, for example, on purely synthetic arguments.

2 Short Construction of the Regular Heptagon

Problem: Given the center O and a vertex $A = A_0$ of a regular heptagon, construct the remaining vertices $A_1, A_2, A_3, A_4, A_5, A_6$.

Solution: See Figure 1. Let point B be symmetric to A with respect to O (three steps). Construct the perpendicular bisector b of BO, and find its intersection H with the circle centered at A, passing through B (equivalently: Fold along a line d through A that carries B onto H on line b: three steps). Construct line a through H, perpendicular to b, and apply procedure (4) to the pair (A, a) and

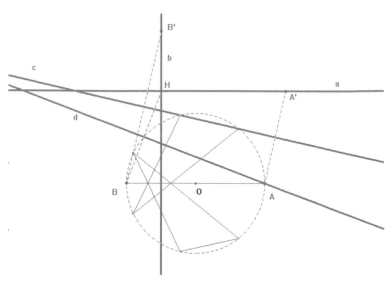

Figure 1. Constructing a regular heptagon.

(B, b) (two steps). Three creases c are found, and they intersect at three points A_1, A_2, A_4. Their final reflection on line AO yields A_6, A_5, A_3 respectively.

Proof. We shall prove that, in a regular heptagon $A_0 A_1 ... A_6$, each of the lines $A_1 A_2, A_2 A_4, A_4 A_1$ reflect A_0 onto line a and B onto line b. Since (4) has no more than three solutions, we can deduce that, conversely, if three lines satisfy (4), they form a triangle inside a regular heptagon. To prove our statement, we shall use complex variables, which is the most convenient analytical tool when dealing with regular polygons. Let us identify a point Z with its complex coordinate $z = x + iy$ and denote its conjugate by $z^* = x - iy$. If we choose vertices $O = 0, A_0 = 1, B = -1$, then it is easy to calculate $H = (-1 + i\sqrt{7})/2$. Let $A_1 = \exp(2\pi i/7) = \cos(2\pi/7) + i \sin(2\pi/7) = \varepsilon, A_2 = \varepsilon^2, ..., A_6 = \varepsilon^6$. From $\varepsilon^7 - 1 = (\varepsilon - 1)(\varepsilon^6 + \varepsilon^5 + ... + \varepsilon + 1) = 0$, we deduce that $u = \varepsilon + \varepsilon^2 + \varepsilon^4, v = \varepsilon^3 + \varepsilon^5 + \varepsilon^6$ are the roots of $x^2 + x + 2$, hence $u + v = -1, u - v = i\sqrt{7}$. As for reflections, it can be easily proved that, given two points z_1, z_2 that lie on the unit circle (so that $z_1^* = z_1^{-1}, z_2^* = z_2^{-1}$), the line through z_1, z_2 reflects point z onto $z_1 + z_2 - z_1 z_2 z^*$. By applying this formula to $z_1 = \varepsilon, z_2 = \varepsilon^2, z = 1$, we find that the line $A_1 A_2$ reflects A_0 onto $\varepsilon + \varepsilon^2 - \varepsilon^3$, whose y-coordinate is $[\varepsilon + \varepsilon^2 - \varepsilon^3 - (\varepsilon^6 + \varepsilon^5 - \varepsilon^4)]/2i = (u - v)/2i = \sqrt{7}/2$. Likewise, for $z = -1$, we find $\varepsilon + \varepsilon^2 + \varepsilon^3$, whose real part is $(u + v)/2 = -1/2$. This proves that line $A_1 A_2$ accomplishes (4). Identical conclusions hold for the other two sides $A_2 A_4, A_4 A_1$, as we wanted. □

3 Distance from a Parabola

Problem: Given a point K and a parabola determined by its focus D and directrix d, construct points P of the parabola for which the distance $|KP|$ is minimal (or maximal).

Solution: See Figure 2. Let point B be symmetric to K with respect to D (three steps). Construct line b through K, perpendicular to d, and apply (4) to the pairs (D, d) and (B, b) to find a crease c. Let the line n through K, perpendicular to c, intersect c at P. (Depending on the position of K with respect to the parabola, we may find only one solution (a minimum) or three (a maximum and two minima)).

Proof. It is well-known that any line c reflecting the focus D onto a point D' on the directrix is tangent to the parabola at a point P, where $D'P$ is perpendicular to d. We shall show that, if c also reflects B onto B' on b, then c is perpendicular to KP, implying that $|KP|$ is extremal. Consider the midpoint $(B + B')/2 = Q$. Since B' lies on b then $QD, B'K, D'P$ are parallel. Thus $D'P$ is perpendicular to d and KP is perpendicular to c, as we wanted. ☐

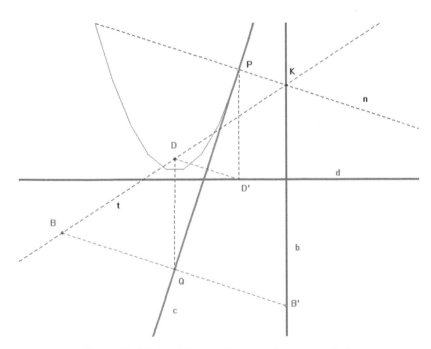

Figure 2. Minimal distance from a point to a parabola.

4   Distance on a Hyperbola

Problem: Given a hyperbola determined by its asymptotes u, v and one of its points R, construct points P of the hyperbola such that the distance $|RP|$ is minimal (or maximal).

Solution: See Figure 3. Let u, v meet at O. First construct the point O^*, symmetric to O with respect to R, then the line u^*, through O^*, parallel to u. Let u^*, v meet at U. Construct the point R^*, symmetric to R with respect to O and the line r, through R^*, parallel to v. Apply (4) to the pair $(U, u), (R, r)$ to construct a line c that reflects U onto a point U' on u and R onto R' on r. Then the line t through U', parallel to c , meets the line RR' at P.

Proof. We shall assume the following result is well-known: If P, R are any two points on a hyperbola, then the line PR intersects the asymptotes at points M, N such that the two midpoints are the same. $(M + N)/2 = (P + R)/2$.

Assume the line RR' intersects u, v at M, N, and let $C = (U + U')/2$. Then $N = (R + R')/2$ lies on c (by the definition of r), and we have $RM = UC = CU' = NP$. This implies $(M + N)/2 = (P + R)/2$, so that, by the previous remark, P lies on the hyperbola which passes through R and has u, v as asymptotes. Let t meet v at S. Then $PS = CN = U'P$. Therefore t intersects the asymptotes at U', S, and their midpoint is $P = (U' + S)/2$.

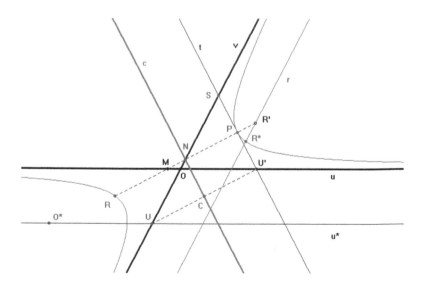

Figure 3. Constructing minimal distances on a hyperbola.

By the same remark, P must count twice within the intersections of t with the hyperbola, and this means that t is tangent at P. Since RP is perpendicular to t, P is extremal, as we wanted.   □

5 The Problem of Philo of Alexandria (or L'Huillier's)

Problem: Given a point Q and two lines u, v intersecting at K, construct the line n through Q that intersects on u, v a segment UV of minimal length.

Solution: See Figure 4. Reflect Q about u to find Q^*. Construct the line q through Q^*, parallel to u. Construct the point K^* symmetric to K with respect to Q. Construct the line k through K^*, parallel to v. If we apply (4) to the pairs $(K, k), (Q, q)$, we always find a unique solution c. Then n is the line through Q, perpendicular to c.

Proof. We could adapt the argument of the previous paragraph (the two problems being strictly related), but we prefer to give an independent proof based on differential arguments.

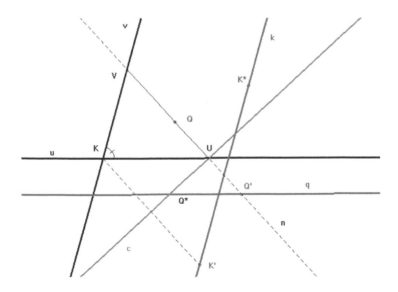

Figure 4. Philo's problem.

Let the lines u, v form an angle θ. Choose a (nonorthogonal) Cartesian reference with $Q = (0,0)$ and xy-axes parallel to u, v, respectively. If $K = (a, b)$, then a line $y = mx$ through Q intersects u, v at $U = (b/m, b)$, $V = (a, ma)$, respectively, and one calculates

$$UV^2 = (b/m - a)^2 + (b - am)^2 - 2(b/m - a)(b - am)\cos\theta,$$

$$\frac{d(UV^2)}{dm} = \frac{2a^2(m - b/a)(m^3 - m^2\cos\theta - m(\cos\theta)b/a - b/a)}{m^3}.$$

(Notice that, for $m = b/a$, the line $y = mx$ intersects u and v at the same point, giving a null segment). We shall show that condition (4) implies that this derivative vanishes, i.e., UV^2 is minimal. Let c be a line which simultaneously reflects Q onto Q' on q and K onto K' on k. Then the midpoint $C = (K + K')/2$ lies on the y-axis, and the midpoint $U = (Q + Q')/2$ lies on u. Also, $\angle KCU$ and $\angle CUQ$ are both right angles. We can therefore calculate CU^2 by applying Pythagoras' theorem in two ways:

$$CU^2 = QC^2 - CU^2 = VK^2 - QU^2 = (b - am)^2 - (\frac{b^2}{m^2} + b^2 - 2(\cos\theta)\frac{b^2}{m})$$

$$= KU^2 - KC^2 = KU^2 - QV^2 = (a - \frac{b}{m})^2 - (a^2 + a^2m^2 - 2a^2m\cos\theta).$$

By subtracting, we find precisely what we wanted: $d(UV^2)/dm = 0$. $\square$

6   Distance on an Ellipse

Problem: Given an ellipse (a hyperbola) by assigning its foci and one of its points R, construct a point P of the ellipse (hyperbola) for which the distance $|RP|$ is minimal (or maximal).

Solution: See Figure 5. Given the two foci E, F of the ellipse E, first construct its axes a, b and center O. Then construct a line t as an (internal) angle bisector of $\angle ERF$. Intersect t with a at U and with b at V. Now construct R^*, symmetric to R with respect to O, then H symmetric to O with respect to R^*. Construct two lines: u through H, perpendicular to b, and v through H, perpendicular to a. Apply (4) to the pairs (U, u) and (V, v). Either one or three creases c are found. The line through R, perpendicular to c, meets c at a point P that is extremal, and c is tangent to E at P.

Note: An identical construction holds for the hyperbola, provided we take as t the external, rather than internal bisector of $\angle ERF$.

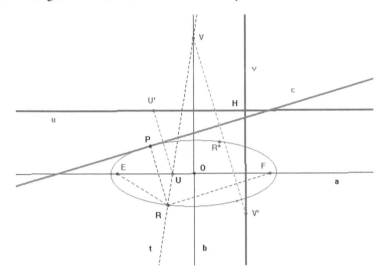

Figure 5. Constructing minimal distances on an ellipse.

Proof. Let $x = a\cos\theta$, $y = b\sin\theta$ represent the ellipse E. The lines tangent and normal to E at $R = R(\theta) = (a\cos\theta, b\sin\theta)$ have equations $x(\cos\theta)/a + y(\sin\theta)/b = 1$, $x(\sin\theta)/b - y(\cos\theta)/a = [a/b - b/a]\cos\theta\sin\theta$, respectively. Given R, we look for points $P = P(\phi)$ on E which are extremal, i.e., such that the vector $RP = (a[\cos\phi - \cos\theta], b[\sin\phi - \sin\theta])$ is perpendicular to the tangent to E at R: $x(\cos\theta)/a + y(\sin\theta)/b = 1$. This condition can be written $a^2\cos\theta/\cos\phi - b^2\sin\theta/\sin\phi = a^2 - b^2$ or, equivalently, $(r - v)[r^3b^2v + r^2(2a^2 - b^2) + r(2a^2 - b^2)v + b^2] = 0$, where the classical substitutions $r = \text{tg}(\phi/2)$, $v = \text{tg}(\theta/2)$ have been used in order to produce polynomials. Besides $r = v$ (yielding $P = R$), one finds either one or three nontrivial solutions, corresponding to a maximum and possibly two minima for $|PR|$.

Let us now describe analytically the suggested construction. The line normal to E at $R = R(\theta)$ intersects the axes at U ($[a/b - b/a]b\cos\theta, 0$) and V ($0, -[a/b - b/a]a\sin\theta$). Define the point $H(\theta) = (-2a\cos\theta, -2b\sin\theta)$, the lines u: $y = -2b\sin\theta$ and v: $x = -2a\cos\theta$, and apply (4) to the pairs (U, u), (V, v). We shall prove that the tangent to E at an extremal point P, as characterized above, is a proper crease. In fact, a tedious but straightforward calculation for U', the image of U under the reflection on the tangent at $P = P(\phi)$, yields the following coordinates:

$$x_{U'} = (a^2 - b^2)\frac{(a^2\sin\phi - b^2\cos\phi)\cos\theta + 2a^2b^2\cos\phi)}{a[a^2\sin\phi + b^2\cos\phi]};$$

$$y_{V'} = -2b \sin \phi [(a^2 - b^2) \cos \phi \cos \theta - a^2]/[a^2 \sin \phi + b^2 \cos \phi].$$

In order to prove that U' lies on u, one must prove that $x_{U'} - x_H = 0$. The vanishing of this difference, in terms of r, v turns out to be equivalent to

$$(r + v)[r^3 b^2 v + r^2 (2a^2 - b^2) + r(2a^2 - b^2)v + b^2] = 0$$

whose nontrivial roots are the same as those of the polynomial we have calculated above. Similar conclusions hold for V, and this concludes the proof. $\square$

7 Alhazen's Problem (Reduced Form)

Alhazen's (general) Problem is the following: Given a circular mirror (billiard) and two arbitrary points K, Z, find points P on the mirror where a ray (ball) KP must reflect (bounce) in order to reach Z. It is well-known (see [2]) that this problem may have four, two, or no solutions, depending on the number of real roots of a polynomial of degree 4. Here we shall study a special case, occurring when one of the given points lies on the mirror. In this case, the problem has a trivial solution so that the relevant algebraic degree reduces to three. Let us state the problem more precisely.

Problem: Given a point K and a circular mirror, determined by its center O and one of its points U, construct points P on the mirror such that a ray UP is reflected onto a ray that hits K.

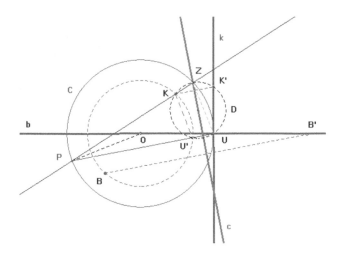

Figure 6. Alhazen's Problem.

Solution: See Figure 6. Let point B be symmetric to K with respect to O. Construct line b through O and U and line k through U perpendicular to b. Apply (4) to the pairs (B, b) and (K, k). Either one or three creases c are found (depending on the mutual positions of O, U, K). A ray n that we seek is perpendicular to c.

Proof. Here we shall use purely synthetic arguments. Let C denote the circle through U centered at O. Let the line through U, perpendicular to c, intersect C in P. Let the line PK intersect C at (P and) Z. Since line UK' is tangent to C at U, the following angles (measured mod π) are equal: $\angle KZU = \angle PZU = \angle PUK' = \angle U'UK' = \angle KU'U$, so that Z, K, K', U, U' lie on the same circle D. Since B' lies on line b which is perpendicular to k, we have $\angle K'UB' = \angle BU'K = \pi/2$ (mod π), so that U' belongs to circle B, whose diameter is KB. As B and C are concentric, D intersects them at two parallel chords KU', ZU. We conclude that both triangles KPU' and ZPU are isosceles, so that radius OP bisects their angle at P, hence UP is reflected by the circular mirror C onto a ray which hits K, as we wanted. □

Note: Alhazen's Problem can be generalized by replacing the circle C with an elliptic (hyperbolic) mirror. The algebraic discussion still leads to a polynomial of degree 3, but finding a reasonably light paper-folding construction for this case is an open problem.

References

[1]   Huzita, H. and Scimemi, B., "The algebra of paper-folding (origami)", *Proceedings of the 1st International Meeting on Origami Science and Technology*, Ferrara (1989), 215-222.

[2]   Neumann, P., "Reflections on reflection in a spherical mirror", *American Mathathematical Monthly*, Vol. 105 (1998), 623-628.

Part Two

Origami Science and Applications

While origami is hundreds of years old and the mathematical sciences are thousands of years old, the two fields have really only begun to interact within the last few decades. As the papers in this section demonstrate, their offspring are many, varied, complex, and fascinating. The information flow in these hybrid investigations goes both ways: The techniques of origami can be brought to bear on real-world problems, often providing unique and efficient solutions, and the tools of mathematics, geometry, and physics can be applied to origami problems, taking the art in new—and sometimes unexpected—directions.

With the advent of high-speed and inexpensive computing power, it becomes possible to imagine, visualize, and/or even fold (in a virtual way) origami more complex than can be folded by hand. Along these lines, Alex Bateman and John Szinger present software tools for origami design, respectively, for tessellations—geometric origami patterns reminiscent of Moorish tilings—and *The Foldinator*, a virtual-reality simulation tool in which the folding takes place in cyberspace.

Are they scientists or artists? Sometimes, a bit of both. Some inhabitants of the boundary between origami art and science have used the techniques of origami to solve real-world structural or design problems. Koryo Miura, a pioneer in the application of origami to space structures, describes how origami techniques elegantly solve an age-old problem: How do you open and close a large map in a confined space? In a similar vein, Tomoko Fuse and coworkers demonstrate how an origami twist pattern leads to a rigid, robust pot—no glue required.

The more geometrically minded within the origami field have long made polyhedra a favorite target genre, which can also be seen in the other two sections of this book. Here, I enumerate a novel class of intersecting polyhedra and provide an origami implementation of them. Miyuki Kawamura demonstrates how trigonometric functions can be implemented with folding, giving examples taken from various polyhedra. Norika Nagata presents a geometric study of a particular origami box, demonstrating quantitatively the relationship between starting crease angles and final box dimensions.

To the origami folder, control is everything: tearing or crumpling is an admission of failure. But to Brian DiDonna, crumpled origami provides an opportunity to analyze the stresses and thermodynamic properties of crumpled paper, giving insights into crumpling in situations where the consequences may be severe. Biruta Kresling, by contrast, shows how origami patterns can represent the stress, strain, and structural integrity of an unusual type of bamboo.

Finally, Ethan Berkove and Jeff Dumont present an analysis of the folding novelty known as the flexagon, one of the first folded paper shapes to receive mathematical scrutiny. Then, we move from the roots of the field to tap into one of the newest areas: the burgeoning field of complexity and self-assembly, as Radhika Nagpal describes a language for self-assembling cellular automata that realize origami forms.

Robert J. Lang

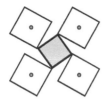

Computer Tools and Algorithms for Origami Tessellation Design

Alex Bateman

1   Introduction

Origami tessellations are a genre of origami that started in the late 1960s with models such as Momatani's stretch wall and later works by Shuzo Fujimoto in his book *Seizo Soru Origami Asobi no Shotai (Creative Invitation to Paper Play)*. More recently, Chris Palmer has brought the art to new heights. The mathematics of the topic have been developed by a number of people including Toshikazu Kawasaki, Helena Verrill, Thomas Hull, and Robert Lang.

To date, there are very few tools available specifically for the design of origami models. The most notable example is TreeMaker by Robert Lang, which has been used to design complex origami bases for constructing insects, animals, and even an Allosaurus skeleton. In this paper, a new program called Tess that can be used to design origami tessellations is presented.

Before the discussion of the program itself, it is necessary to introduce the concepts of crease pattern, folded pattern, and light pattern, A crease pattern is the set of all creases that can be used to fold a piece of paper into a flat-foldable origami tessellation. An example crease pattern is shown in Figure 1(a). A folded pattern shows the folded origami tessellation, where the paper is completely transparent and only the crease lines are shown. This represents an approximation of what is seen when holding the folded crease pattern up to a bright light. This representation is a computational construction that is useful because one does not need to consider the layers of the paper; see Figure 1(b) for

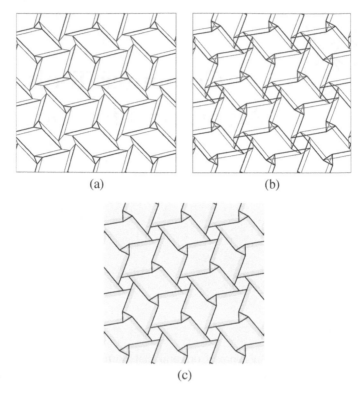

Figure 1. (a) An origami crease pattern. (b) A folded pattern. (c) A light pattern.

an example. The final pattern that we are considering is the light pattern, which shows what an origami tessellation made from semitransparent paper looks like when held up to light. For an example of a light pattern, see Figure 1(c).

2   Algorithms

The key algorithm used in origami tessellations transforms a tiling of the plane into a flat-foldable crease pattern. This algorithm has been developed and used by people in the field such as Paulo Barreto and Chris Palmer, and the algorithm presented here is based on their work. I will describe this algorithm including the parameters used to enumerate the possible crease patterns. This algorithm and the correct choice of parameters leads to the novel result that it can also generate a representation of the folded pattern. Given the folded representation, I present an algorithm that can generate the light pattern of the origami tessellation.

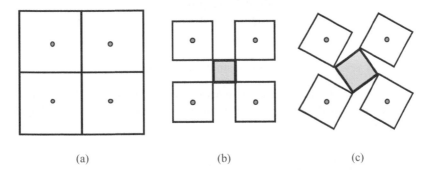

(a)   (b)   (c)

Figure 2. The three steps used to generate an origami crease pattern from a tiling. (a) Generate tiling; the filled dots are the origin of scaling (derived from the vertices of the orthogonal dual of the tiling) and rotation in Steps 2 and 3. The filled dots are linked for a dual of the original tiling. (b) Scale polygons of tiling, and generate a new "baby" polygon, shown shaded. (c) Rotate polygons to generate final pattern. The "baby" polygon vertices do not need to be explicitly calculated as they coincide with the vertices of the original tiling.

2.1   Crease Pattern Generation

Origami tessellations are generated from a tiling of the plane. This tiling must have a dual such that every edge in the tiling is orthogonal to its corresponding edge in the dual. This dual tiling is called the orthogonal dual. Figure 2 shows the steps involved in creating an origami tessellation crease pattern from a tiling for the archimedean tiling 4^4. Given a tiling, each tile must be scaled and rotated, and the origin of scaling and rotation for each tile is defined by the location of the corresponding vertex of the orthogonal dual. In the algorithm presented here, all tiles are rotated and scaled by a uniform amount; however, this is not a requirement to make an origami tessellation. Different tiles in the tiling can be scaled and rotated by different amounts, but this makes the mathematics more complex. During the scaling step, new baby polygons are produced. These polygons are formed by joining the vertices of tiles that are adjacent in the original tiling. The natural choice of parameters would seem to be the scale factor (α) and rotation angle of each tile (θ). However, I use the ratio of the lengths of the tile compared with the length of the edge of the baby tile (α/γ) and the pleat angle (ϕ) (see Figure 3). There is a simple relationship between these two parameter sets, shown in the equations below:

$$\theta = \arctan\left(\frac{1}{\tan\phi + \left(\frac{\alpha/\gamma}{x\cos\phi}\right)}\right);$$

$$x = \frac{2\sin(\pi/n_1)\sin(\pi/n_2)}{\sin(\pi/n_1 + \pi/n_2)};$$

$$\alpha = \frac{1}{\cos\theta + \sin\theta\tan(\theta + \phi)}.$$

Here n_1 and n_2 are the number of sides of the polygons at each end of the pleat being described by the parameters under discussion. So, for the tiling in Figure 2, n_1 and n_2 equal 4.

2.2 Folded Pattern Generation

The same algorithm that is used to generate the crease pattern of a tiling can be used to generate the folded pattern. The choice of parameters discussed above allows for a simple relationship between the parameters for the crease pattern and the folded pattern. To generate the folded pattern, we simply negate the pleat angle. Thus, for a crease pattern with a pleat ratio of 25 degrees, we can generate the folded pattern by using a value of −25 degrees. This result is rather surprising, but it allows us to view origami tessellations as they will look once folded. This is a great aid in design of these tessellation patterns.

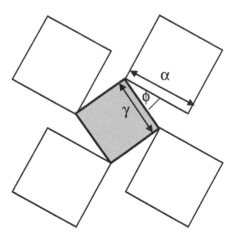

Figure 3. The parameters used to specify origami tessellations. Once a tiling has been chosen, two parameters can be used to describe the simplest type of origami tessellation. The parameters used are the pleat angle, called ϕ, and the pleat ratio (α/γ). α is also the scale factor by which the original tile has been scaled. Thus, an α of 0.5 means that is half the size of that in the original tiling.

2.3   Light Pattern Generation

The conceptually simplest algorithm to generate a light pattern is to consider each pixel in the folded representation and count the number of polygons in which it lies. (This is equivalent to finding the number of layers of paper at any point.) This value is then used to assign a grayscale value to the pixel. For pixels with one layer, a light gray is assigned, and darker grays for thicker regions. This algorithm requires a comparison for every pixel in the pattern. If n is the number of pixels in one dimension, then the complexity of this algorithm is $O(n^2)$. This means that doubling the resolution causes the time needed to calculate the light pattern to quadruple. In contrast, I have implemented an algorithm that requires only $O(n)$ computations. This algorithm considers one horizontal line in the pattern at a time. For polygons that intersect this line, we know that at the leftmost intersection, we need to increment the number of layers, and at the rightmost intersection point, we should decrement the number of layers. This algorithm will calculate a high-resolution light pattern in less than a minute.

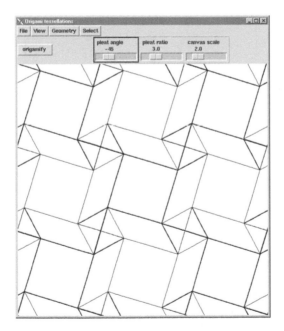

Figure 4. A screen shot from the GUI version of the Tess program. Note that the pleat angle is −45, and therefore shows a folded pattern.

3   The Tess Program

The algorithms described in the previous section have been implemented in a computer program called Tess, which has been written using object-based PERL. The software is freely available and can be downloaded on the web at: http://www.sanger.ac.uk/Users/agb/Origami/.

The Tess program can be used in two different ways: A command line interface can be used, or, if PerlTK is installed, a Graphical User Interface (GUI)-driven version can be used (see Figure 4). The object model used in the program is represented in Figure 5.

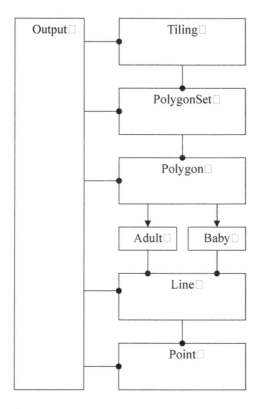

Figure 5. The object model of the Tess program. Container objects are joined to other objects by a line with a circle at its end. Inheritence relationships are shown with arrows. A tiling object contains a collection of polygonset objects, for example. Adult and baby objects are subclasses of the polygon object. The output object is a list of objects to be rendered.

4   Conclusions

I have shown that an algorithm developed by others can also be used to generate representations of the folded origami. This can either show the crease lines or the light pattern of the tessellation. The use of these algorithms in the Tess program allows the design of complex new origami tessellations in silicon, moving the origami practitioner a step closer to the paperless office.

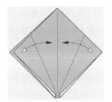

The Foldinator Modeler and Document Generator

John Szinger

1 Introduction

Foldinator is a software application for visualizing objects made by the sequential folding of flat sheet materials and for generating annotated documents that record and display the sequence of steps in the object's creation, providing a useful set of instructions to others. Such sequences can be used to represent origami models and other products. In contrast to a generic two-dimensional drawing or three-dimensional CAD program, *Foldinator* models the special qualities of sheet material, particularly its geometric configuration with respect to the constraints of manipulating two-dimensional materials in three-dimensional space.

As a software development project, the goal of *Foldinator* is to create a tool to intuitively explore origami folding and document origami models in a sharable format. *Foldinator 1.0* allows the sharing and display of documents over the World Wide Web and has a downloadable tool for creating *Foldinator* origami models and documents. Future versions of *Foldinator* will feature high-resolution graphic output suitable for print publication. I intend to make *Foldinator* publicly available as a shareware application to serve as a resource to the origami community and promote online sharing of origami models.

Hand-drawing origami diagrams requires specialized tools and skills, and is a tedious process, especially for complicated models. Since it involves much repetition and variation, the process seems to be a good candidate for an automated, computerized approach. Currently, someone wishing to use a computer to create origami diagrams has several options. One of these is a two-dimensional drawing and illustration program such as *Illustrator* or *Freehand*. The main drawback of these programs is their lack of a geometric model for the paper and the operations performed. The author is required to manually construct all of this information in his drawings, at a level that simply emulates the appearance but not the underlying structure. Another alternative is to use a three-dimensional modeling and animation package. These tools allow the author to create a model representing the origami, but they have drawbacks of their own. They tend to be expensive and complicated, requiring specialized skills and experience. They employ generalized modeling systems, so many of the capabilities of the software are extraneous to the task at hand. Also, the renderers used in these packages are not designed for origami models.

My solution was to develop a custom application for modeling origami. Origami is a formal system, well-suited for software representation. Creating diagrams is a good application for an authoring tool or expert system because the geometry of paper obeys a relatively small set of well-understood constraints. Similarly, the vocabulary of paper-folding includes a small set of operations, each of which can be represented in a computer simulation. The depth of the experience comes from the ability to repeat operations and to choose what operation to apply next at every step of the way. Indeed, the same qualities that make origami an elegant, aesthetic, and appealing art form in the real world are the ones that make it an interesting process to program as a computer simulation.

2   *Foldinator* Design and Development

Foldinator has three primary design goals: Users can model an origami model on a computer; users save annotated steps to generate diagrams to share with others; and users interact with an intuitive UI (User Interface) based on traditional origami symbols when manipulating the model.

I am developing *Foldinator* using a combination of *Macromedia Director* and *Macromedia Flash*. *Director*, the main development environment, is well-suited for building an interactive UI and for rapid prototyping. It has high-level structures to represent and manage collections of visual and data objects, and includes an object-oriented programming language, Lingo, that is well-suited for representing models and geometry. Additionally, *Director* has a powerful graphics management and display engine, which greatly simplifies the task of programming the visual display of the interface and of rendering the origami model. *Foldinator* incorporates parts of a Lingo code library, *Dave's 3D Engine*, which handles the low-level three-dimensional geometry implementation and is publicly available, open-source shareware. *Flash* is designed for the flexible display of resolution-independent, vector-based graphic objects. *Flash* elements can be freely scaled, rotated, distorted, and otherwise manipulated without introducing visual artifacts. *Director* can then encapsulate *Flash* objects. I created my two-dimensional primitives of the three-dimensional model in *Flash*.

These development tools permit the resulting application to be available to the widest possible user base. The *Foldinator* authoring tool will be a downloadable application for the PC and Macintosh. The *Foldinator* document viewer will run inside any web browser as a Shockwave document. *Foldinator* origami model documents are based on the .TXT format and accessible to authors and viewers on any platform.

At the heart of the *Foldinator* application is an object called Paper. Paper begins its life as an unfolded square of paper with two sides and (nominally) zero thickness. When the user creates a fold, the crease divides the square polygon into two smaller polygons joined with a common, hinged edge. As the model progresses in its development, new folds propagate through multiple layers of paper, creating numerous new polygons in the model, according to the type of fold. Additionally, clarity of display necessitates variation from "true" three-dimensional projection. Rather, aligned layers of paper are shown systematically offset so that the viewer can read the diagram. This requires modification of the three-dimensional engine to create the projection from the modeler to the renderer. Special attention was given to lines and line weights in the development of the model view. The language of origami diagramming requires subtle and meaningful distinction between various kinds of lines, such as creases and edges, and the various kinds of dashed and dotted lines. In this regard, the use of *Flash* has paid off handsomely.

The Paper exists in a Scene. The Scene is a composite 2D/3D object that also contains diagramming symbols and text annotations. The Sequence contains a series of Scenes and represents the origami model as a work in progress or as the completed set of folding instructions. It also includes global annotations such as title, author, and global annotations. A Sequence can be saved or loaded as a file. This is a *Foldinator* document. Currently the Sequence is encoded as a .TXT file. Rendered scenes are encoded as .BMP or .JPG files and are referenced by the .TXT file. Support for .EPS or similar print-resolution images is planned for a future version of *Foldinator*.

3   Using *Foldinator*

Now it is time for a guided tour of the *Foldinator* application. There are three Scene views of the origami model: the Main view, the Map view and the

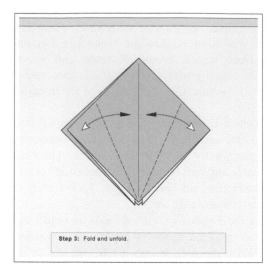

Figure 1. Main view. Note the visual offset of aligned layers of paper.

Section view. Additionally, a Sequence view provides a high-level view of a series of diagrams in thumbnail format. Within each view are various controls, tools, and other UI elements. The Main view is designed to look and act as much like an origami model as possible. It is manipulable and annotatable. The Map view shows flat paper with all the folds in place. This is sometimes called the fold pattern. The Section view displays a cross section of the model to show the arrangement of the various layers of paper. The Sequence view shows a series of thumbnails of the folding steps. The user may use this to step through his model sequentially.

A new *Foldinator* document presents the paper in the center of the screen and the various controls arrayed around it. At the top of the window is the Menu Bar, where the user can access various file- and application-level commands. On the left side is the Tool Palette. The tools use the icons of traditional origami notation, so their functions should be apparent to anyone familiar with reading origami diagrams. They include various types of folds, plus Rotate, Flip, and Zoom functions. At the bottom of the window is the Annotation space, where the user can type in text instructions to accompany the symbolic markup of the step. The Step Controls (lower right) allow the user to execute the current step and move ahead to the next one, and also to move backward and forward through the sequence of steps.

Often a user will want to set the Paper properties before beginning to fold the model. Access to this functionality is provided through the Menu Bar or by

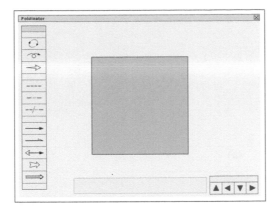

Figure 2. A new *Foldinator* document.

right-clicking on the Paper itself. Paper properties include the shape or aspect ratio of the paper, as well as its color or graphic pattern, and its thickness. (The default values for paper shape and color are: a square sheet with one white side and one gray side. Currently only "ideal" paper is supported. Ideal paper has zero thickness (although it cannot pass through itself or do other "impossible" things), absolute stiffness, always produces precise and permanent creases, and cannot be stretched, curved, cut or torn, only folded.

4 Making a Fold

To begin, the user selects the type of fold. Currently the available choices are valley, mountain, and reverse folds. Support for more complex folds, such as petal fold, squash, rabbit ear, and sink are planned, but most of these can be accomplished by sequences of simpler folds. *Foldinator* supports Snap To by default, although it can be turned off. Most origami folds, especially at the beginning of a model, are placed relative to existing features of the geometry of the paper. The user can easily place a fold at the bisector of two existing lines, such as parallel or adjacent edges of the paper, or an edge and an existing crease, or two existing creases. The user can also snap to a point at the intersection of any pair of creases or the intersection of a crease and an edge, both for locating the end point of a crease or for locating the destination of a locatable point on a folded flap.

To make a fold, simply select the type of fold and drag a line across the model by placing one endpoint, then the other. Once the fold has been located, the direction of the arrow must be specified; that is, the user must indicate which side of the model will be folded onto the other. When the second

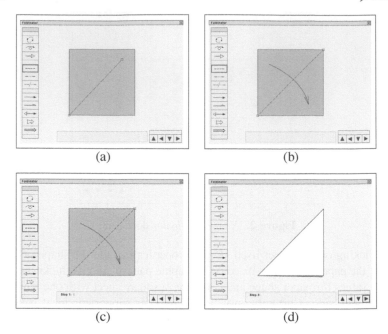

Figure 3. (a) Dragging a line to make a valley fold. (b) Orienting the arrow. (c) Annotating the step. (d) Executing the step.

endpoint of the fold is placed, the arrow automatically appears, with the arrow head on the same side of the line as the mouse. To change the direction of the arrow, the user simply drags the mouse across the line and clicks. Once the fold is placed, the user may annotate the step. This is accomplished simply by clicking in the annotation field and typing.

All that remains is to execute the step using the Step Controls in the lower right of the application. When the user clicks on the Next icon, the model animates the action specified and updates its internal state for the next step. The Step Controls allow the user to pop up to the Sequence view, to step down back into the Scene view, and to move backward or forward through the various steps in the model sequence.

5   Conclusion

Several issues emerged during the development of *Foldinator*. Foremost was the look and feel of the application. The lines should appear smooth and free of aliasing artifacts. Similarly, the projection from the three-dimensional model to the two-dimensional image had to elucidate the features of the model that were important to folders. These concerns influenced my choices for the development platform. The complexity of elaborate origami models in three-dimensional may turn out to be another issue. As it is still under development, I have not yet tested *Foldinator* with very complex models. While I believe that my constraints are sufficient to prevent the paper from folding through itself or doing other "impossible" things, I have not proven in a rigorous mathematical sense that this is, in fact, the case.

Some aspects of an origami simulation are more difficult to implement, and I am deferring them to a future version of *Foldinator*. In particular, I would like to support curvature in folds and in the paper itself to allow a sculptural approach, especially in the finishing stages of a model. This will require me to introduce splines to the three-dimensional model, and will require a reworking of significant amounts of code. I may update my codebase to use *Director 8.5*, which includes mesh-based three-dimensional geometry objects.

Foldinator is currently under development by Zing Man Productions. You can become a beta tester by sending email to: john@zingman.com. Please note that the features described in this paper do not constitute a formal specification, and may change in the final release version. If public response to *Foldinator 1.0* is sufficiently positive, I will continue to add features to create more so-phisticated versions of the software. Planned features include support for paper with nonzero thickness and other variable physical qualities, support for paper of arbitrary nonrectangular shape, support for curved folds and curved paper, support for mapping images onto the paper as a texture map, and support for print-quality rendering of the steps. Based on the initial feedback I have gath-ered from fellow origami enthusiasts, I expect that *Foldinator* will fill a useful role in the community of paper-folders who wish to share their creations with one another. I look forward to its completion and public release.

The Application of Origami Science to Map and Atlas Design

Koryo Miura

1 Introduction

After its invention in ancient China, paper has been the main media for maps and atlases. Although electronic media is becoming increasingly popular for maps, the popularity of paper maps will not be changed substantially in the future. Therefore, designing packaging forms for paper maps and atlases remains one of the major concerns for map publishers. The packaging of paper is affected by its intrinsic properties. Among these, the nearly inextensional property is the major constraint posed on the packaging design. In mathematical terms, the Gaussian curvature at any point on the surface of a paper should be zero as well as invariant for any transformation of the form. The subject has been studied in more detail as the theory of flat origami. This principle can be a constraint as well as a guideline for design. This paper presents two different types of packaging of geographical information guided by this principle. One is an easy-to-fold/unfold map based on the result obtained from space engineering research, and the other is a novel atlas design solving the problem of discontinuity at the page borders.

2 Easy-to-Fold/Unfold Map Design

The packaging of a large expanse of thin planar material into a smaller volume is the current major problem of large space structures. Using tools such as solar power satellites, solar sails, and large antennas would not be possible without solving that packaging problem. In 1970, we presented a particular surface called "the developable double corrugation surface" (DDC surface) [Miura, 1970]. It is the abstract surface obtained from contracting an infinite plate bi-axially within the plane of the plate. It should be noted that the surface fulfills the geometric constraint mentioned above. This proposition was proved later by numerical study [Tanizawa and Miura, 1978]. The particular properties of this corrugated surface prompted us to apply it to the packaging and deployment of large space structures [Miura, 1980]. At the same time, the application of this concept to map design was studied [Miura and Sakamaki, 1977]. The "Map of Venice", based on the concept, was published with the support of Olivetti Corp, Japan [Miura, 1978] and the result was reported at the ICA Conference in Tokyo [Miura *et al*, 1980]. This type of map, even though it demonstrates favorable properties as a folded map, has not been used widely because of the difficulty in machine-folding. Recently, a method of mass production became possible to meet commercial needs. It seems adequate to report a summary of the design principles and properties of the concept.

2.1   Geometry of the Developable Double Corrugation (DDC) Surface

The Developable Double Corrugation (DDC) surface is generally characterized as the repetition of a fundamental region consisting of four identical parallelograms as shown in Figure 1.

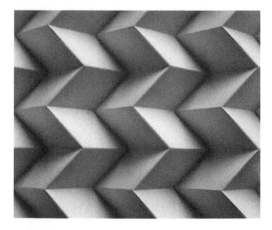

Figure 1. The Developable Double Corrugation (DDC) surface.

The DDC surface is a generalized surface which includes various shapes depending on its parameters. The limited ranges of parameters of the DDC surface are adequate for folded map design. The most influential parameter is the zigzag angle, which is the offset angle from the otherwise straight fold lines. In general, if the angle is large, the packaged volume is not small. On the contrary, if the angle is small, say 1 degree, the folding/unfolding process is not smooth. The appropriate range of the angle is from 2 to 6 degrees. This type of folding is called *Miura-ori* in the map design and origami worlds [Anonym, 1981]. (In the following, the word *Miura-ori* is used in place of generalized DDC surface.)

2.2   Behavior of the *Miura-ori* Map

A typical design example for approximately A2 size paper is shown in Figure 2.

- Automatic folding/unfolding capability.

The *Miura-ori* map is easily unfolded by pulling two opposite diagonal corners. Pushing the diagonal corners together will recollapse the map. Both processes are almost automatic. This behavior is due to the fact that the map's mechanism has a single degree of freedom. This is the reason that concept has been used for the space solar array structures.

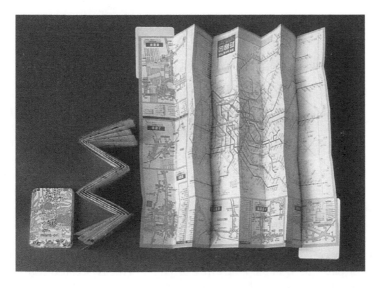

Figure 2. Miura-ori map for Tokyo Transit (courtesy of ORUPA, Ltd.)

- Preservation of sign of folds.

In terms of origami science, the mountain folds and the valley folds are called positive and negative folds, respectively. An arrangement of the fold signs, therefore, completely defines the shape of an origami work. The arrangement of fold signs for *Miura-ori* is unique, and the movement of every fold is linked. Since it has the mechanism of a single degree of freedom, the sign of every fold is preserved. There is no possibility of missing the sign of any fold. On the contrary, for a conventional, normally folded map, there are always two ways per fold, thus the total number of ways to fold the map is very large. Folding the map incorrectly is the major cause of fracture of the paper at folds and nodes.

- Strength of folds.

Another fact relating to the strength of the present map is due to the offset of nodes. As can be expected, when the map is completely folded, the nodes are always offset because of the offset angle. This design feature greatly reduces strain in the vicinity of nodes. In conventional maps, the nodes are stacked at a single point, causing great strain in the vicinity of nodes and indicating another weakness of conventionally folded maps. The preservation of the fold signs as well as the offset of nodes contributes greatly to the durability of the *Miura-ori* map. Because of this property, it does not require strengthening of the paper, such as plastic lamination.

3   Atlas Design—Page Border Discontinuity

3.1   The North-South Problem

The atlas discussed in this paper is the type of road atlas covering a continuous range of an area. Since such an atlas covers a two-dimensional space, the address for any orthogonally divided area should be expressed in the matrix form (m, n), as shown in Figure 3. The west-to-east pages are arranged in "rows," while the north-to-south pages are arranged in "columns". The "window" indicated by a pair of rectangles in the figure is a spread of the atlas, where the needed matrix element should appear. The annoying property of conventional atlas design is the break of continuity of two-dimensional information at the north and south borders of pages, causing difficulties when one wants to go south or north crossing the border of the current page. In Figure 3, if one goes from page (2, 4) to the geographically adjacent page (3, 4), one has to search for page (3, 4) in the long array of pages. This is because conventional atlases are almost exclusively arranged in "rows" only. Such maps consist of joining strips of row elements to make a long single strip, the one-dimensional

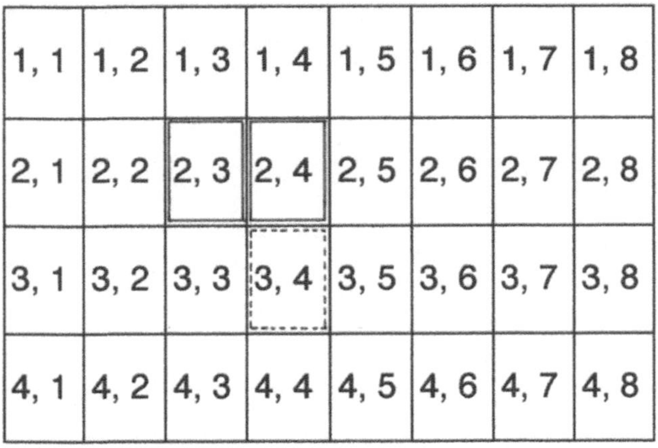

1, 1	1, 2	1, 3	1, 4	1, 5	1, 6	1, 7	1, 8
2, 1	2, 2	2, 3	2, 4	2, 5	2, 6	2, 7	2, 8
3, 1	3, 2	3, 3	3, 4	3, 5	3, 6	3, 7	3, 8
4, 1	4, 2	4, 3	4, 4	4, 5	4, 6	4, 7	4, 8

Figure 3. The north-south problem of the atlas expressed in matrix form ($m = 4$, $n = 8$).

form. Without an excellent memory of figures, most likely one will lose the memory of the previous page. In this paper, the problem of the "southward or northward trip" along the columns is called the *north-south problem*.

The solution of the north-south problem seems to be impossible to obtain because of its strict geometric constraints. With the help of origami science, we found that the present atlas problem could be converted into the map-folding problem.

3.2 Conversion to the North-South Problem of Atlases

In Figure 3, the rectangular lattice represents the abstract matrix expressing the area covered by an atlas, and thus it does not represent any physical object, that is, a map. However, if it is considered as a real map, the present problem can be transferred to the north-south problem of a map, provided it is only to be locally unfolded. After a process of trial and error, we find that it is impossible to solve the problem under the geometric constraints, that is, the Gaussian curvature at any part of the paper surface must be zero throughout the folding/unfolding process. It seems that the only way to overcome this difficulty is to introduce cut lines to avoid the constraint without violating the invariancy of the Gaussian curvature.

Figure 4 shows the fold and cut lines of a rectangular sheet of paper. The mountain lines, the valley lines, and the cut lines are indicated with solid, broken, and double broken lines, respectively. Figure 5 shows the resultant model following this design.

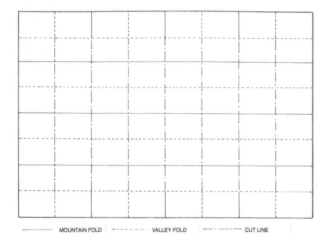

Figure 4. An arrangement of fold/cut lines solving the north-south problem of map $(m = 4, n = 8)$.

The principal features of the model are

1. It consists of several blocks.

2. A bundle of pages are fixed on both sides of each block.

3. Each page is divided by a horizontal valley fold.

4. Each page is double-woven.

5. Each page is opened/closed vertically.

6. Left and right pages are independent.

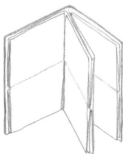

Figure 5. Map model for solving the north-south problem $(m = 3, n = 4)$.

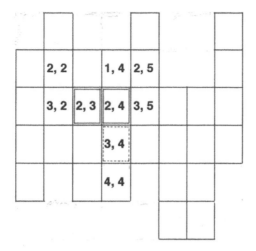

Figure 6. Function of the map model for solving the north-south problem ($m = 4$, $n = 8$).

With this construction, a sheet of rectangular paper turns into a mechanism, which is similar to a well-known game. The function of the map can easily be understood by means of a scheme as shown in Figure 6.

In this scheme, a specific row and a column in the matrix identifies each page. Relative to the fixed window, each page has the ability to scan in lateral (rows) as well as vertical (columns) directions. Let us assume that, at first, the pages (2, 3) and (2, 4) are in the window as indicated in Figure 6. A north-south trip from page (2, 4) to page (3, 4) on the map is simply to turn a page in the vertical direction. Thus, it is clear that the north-south problem of the map is solved. (This solution can be readily applied, with some modifications, to practical design of particular maps. However, the present purpose of this paper is to solve the north-south problem of the atlas and not the map.)

3.3 The Atlas Design Solving the North-South Problem

If we carefully inspect the block of vertical pages of the model, we know that each sheet is formed into a double-woven construction. (The inside pages are dummies and useless.) This construction is due to the process started from a continuous sheet of paper printed on only one side. If we neglect the above process and look at the model topologically, the double-woven pages can be replaced with a single sheet printed on both sides. The result is surprisingly simple. It consists of several booklets bound laterally, where each booklet is arranged vertically as shown in Figure 7.

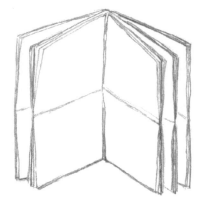

Figure 7. The atlas design solving the north-south problem ($m = 5, n = 8$).

Some design principles are

1. Relatively stiff base sheets should be used for the blocks.

2. A ring-type binding is recommended.

3. The function is similarly explained by Figure 6.

4. Some marks indicating the vertical (column) page number are necessary.

There are no obstacles in engineering a production process for this type of atlas. It is interesting to note that the final result is not a kind of origami work. However, the concept was obtained by transferring the problem to one of origami. Thus the solution of the north-south problem of the atlas was completed.

4   Conclusion

We presented two different ways of packaging geographical information. One is an easy-to-fold/unfold map, and the other is an atlas solving the problem of discontinuity at the page borders.

References

[1] Anonym, "The *Miura-ori* map", *British Origami*, Vol. 88, (1981) 3–5.

[2] Miura, K., "Proposition of pseudo-cylindrical concave polyhedral shells", *IASS Symposium on Folded Plates and Prismatic Structures (Section I)*, Vienna, Austria (1970).

[3] Miura, K. and Sakamaki, M., "A new method of map folding", *Map: Journal of the Japan Cartographers Association*, Vol. 15, No. 4, (1977) 7–13.

[4] Miura, K., "Method of packaging and deployment of large membranes in space", 31st Congress of the International Astronautical Federation, Tokyo (1980).

[5] Tanizawa, K and Miura, K., "Large displacement configurations of bi-axially compressed infinite plate", *Trans. Japan Society of Aeronautics and Space Science*, Vol. 20, (1978) 177–187.

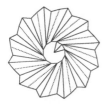

Origami Pots

Tomoko Fuse, Akira Nagashima,
Yasuhiro Ohara, and Hiroshi Okumura

1 Introduction

Though styrene pots are very popular for products such as noodles and soup, they are considered to be a serious cause of environmental pollution. Recently, several kinds of paper pots have been introduced as replacements for styrene items, but they generally consist of several pieces of paper held together with adhesives. While there is a one-piece paper pot, which was patented in the United States in the 1930s [1], this pot also requires the application of adhesives (Figure 1 shows its pattern of creases). We have finally succeeded in making a paper pot from a single piece of paper that uses no adhesive (see Figure 2). These pots can serve as paper containers for a variety of uses. Their design is based upon mathematical and engineering principles, and they can be made on a production line. In this paper, we will demonstrate how to achieve the crease pattern for our pot for given dimensions.

2 Fundamental Property

Let us assume that the bottom of our pot is a regular n-gon with circumradius r_1, center O, height h, and r_2 is the radius of the top. One of the external creases on the side surface meets the bottom and the top at C and D, respectively, and F is the foot of the perpendicular from D to the plane where

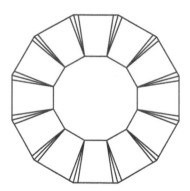

Figure 1. Old pot.

Figure 2. New pot.

the bottom lies, and $\psi = \angle COF$ (see Figure 3). (Paper pots with the crease pattern in Figure 1 correspond to the case in which $\psi = 0$.)

We consider the side surface of the pot to be a part of a circular cone. Precisely speaking, this is incorrect since the bottom is not a true circle, but a regular n-gon. However, our experiments show that we can calculate the dimensions of the crease pattern of our pot with sufficient precision using this assumption in cases when n is large enough. (For example, we can get sufficient dimension in the case where $n = 12$).

Figure 4 shows a side view, and Figure 5 shows the pot unfolded. If we fold it, CD and CE are overlaid, that is, CE is the reflection of CD in CH,

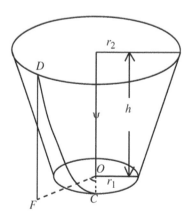

Figure 3. Normal view.

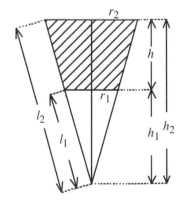

Figure 4. Side view.

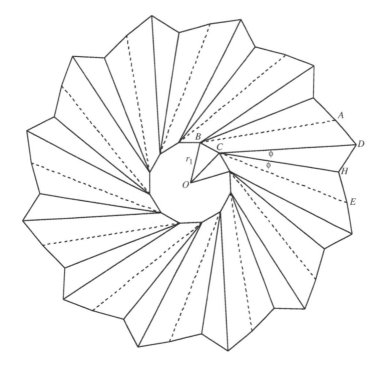

Figure 5. Crease pattern.

where CE is not a crease (note that the dotted lines are not creases; they show the places where the creases are overlaid). Since AD, HD, and HE are arcs (with a common radii), we describe them as arc AD, arc HD, arc HE.

If we regard the sidepiece as one sheet of paper, and cut it off at the generatrix through B, we get Figure 6. It shows that the center T (the vertex of the cone) of the arc AD is the intersection of the perpendicular bisectors of the segments AD and BC. From similar figures in Figure 4, we get $h_1 = r_1 h/(r_2 - r_1)$. The length of the generatrix of the smaller cone is

$$l_1 = TB = \sqrt{h_1^2 + r_1^2}.$$

The length of the generatrix of the larger cone is

$$l_2 = TA = (r_2/r_1)\sqrt{h_1^2 + r_1^2}.$$

Note that these values are determined uniquely by r_1, r_2 and h.

Let $\angle DCH = \angle ECH = \phi$ in Figure 5.

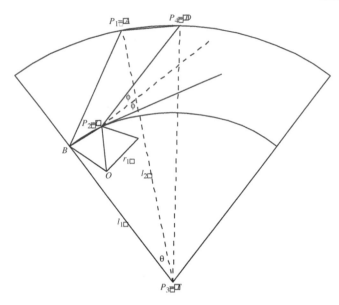

Figure 6. A cut-out side piece of the folded pot, viewed as a cone.

Theorem. ϕ *is independent of* ψ, *and is uniquely determined by* n, r_1, r_2, h.

Proof. In the quadrangle $ABCD$, we abbreviate $\angle DAB$ as $\angle A$. $\angle B$, $\angle C$, $\angle D$ are defined in the same manner. Since the sum of the angles around C is 2π in Figure 5, we have

$$\angle B + \angle C + 2\angle OBC + 2\phi = 2\pi.$$

But $\angle B + \angle C = 2\pi - \angle A - \angle D$, and we have

$$2\phi = \angle A + \angle D - 2\angle OBC.$$

While $\angle A + \angle D = 2\angle TAD$ in Figure 6, we get

$$\phi = \angle TAD - \angle OBC.$$

Hence ϕ is the difference of the base angles of the isosceles triangles TAD and OBC, which are determined uniquely by n, r_1, r_2, h. Therefore, ϕ is independent of ψ and is determined uniquely by n, r_1, r_2, h. $\qquad\square$

By the theorem, it is sufficient to give the parameter ψ along with n, r_1, r_2, h to determine the crease pattern.

3   Uniform Condition of the Top Edge

In general, A and H are not overlaid when we fold the crease pattern. The top edge consists of one layer and three layers, or three layers and five layers in general, and the situation depends on the value ψ. But if A and H are overlaid, all the parts of the top edge consist of three layers. This case may be preferable, and we will show how to obtain such a condition. This condition is designated by special values of ψ. The necessary and sufficient condition which implies the required case is that the equation $\angle ACD = \phi$ holds. Let $\theta = \angle BTA$. We shall get the value θ at first which satisfies this equation. With this value θ, we can get the shape of the quadrangle $ABCD$ in Figures 5 and 6, and we can get the crease pattern. Since $r_2\psi = l_2\theta$, the value of ψ is obtained from $\psi = l_2\theta/r_2$.

Let $l = AB = CD$, and let us relabel as $x = AC$, $A = P_1$, $C = P_2$, $T = P_3$, $D = P_4$, and $d_{ij} = P_iP_j$. Since $\angle ATD = 2\pi r_2/(nl_2)$, $AD = 2l_2\sin(\pi r_2/(nl_2))$. Then $d_{12} = AC = x$, $d_{13} = l_2$, $d_{14} = 2l_2\sin(\pi r_2/(nl_2))$, $d_{23} = l_1$, $d_{24} = l$, $d_{34} = l_2$. Now we regard x and θ as indeterminates.

Note that

$$d_{24} = l = \sqrt{l_1^2 + l_2^2 - 2l_1l_2\cos\theta}$$

has only one indeterminate θ. All the d_{ij}, excluding d_{12} and d_{24}, are determined by n, r_1, r_2 and h. Let us define

$$M = \begin{pmatrix} 0 & d_{12}^2 & d_{13}^2 & d_{14}^2 & 1 \\ d_{12}^2 & 0 & d_{23}^2 & d_{24}^2 & 1 \\ d_{13}^2 & d_{23}^2 & 0 & d_{34}^2 & 1 \\ d_{14}^2 & d_{24}^2 & d_{34}^2 & 0 & 1 \\ 1 & 1 & 1 & 1 & 0 \end{pmatrix}.$$

Since P_1, P_2, P_3, P_4 are points in the plane, we get

$$\det(M) = 0. \tag{1}$$

Equation (1) contains both x and θ. The condition $\angle ACD = \phi$ is expressed as

$$\frac{l^2 + x^2 - AD^2}{2lx} = \cos\phi.$$

While

$$\phi = \angle TAD - \angle OBC = (\frac{1}{2} - \frac{r_2}{nl_2})\pi - (\frac{1}{2} - \frac{1}{n})\pi = (1 - \frac{r_2}{l_2})\frac{\pi}{n},$$

we get

$$\frac{l^2 + x^2 - \left(2l_2\sin(\pi r_2/(nl_2))\right)^2}{2lx} = \cos\left((1 - \frac{r_2}{l_2})\frac{\pi}{n}\right). \tag{2}$$

Equation (2) also contains both x and θ. From Equations (1) and (2) we can get the value of θ. Using *Mathematica* we can obtain the value of θ in a few seconds.

References

[1] "Paper container and method for manufacturing the same", U.S.A. patent No. 6,237,845.

Polypolyhedra in Origami

Robert J. Lang

Among the different genres of the art of origami, one alluring to many is modular origami: origami formed by joining many identically folded units into a geometric shape—typically, some form of polyhedron. There are at least as many modular origami designs as there are polyhedra—actually, there are many more, since a given polyhedron may be folded in many different ways. In most cases, the modular origami designer starts with a polyhedron in mind and develops an origami rendition of it. On rare and happy occasions, the progression goes the other way. So it was with an investigation I made into a particular class of origami modulars—intersecting polyhedra—wherein the origami investigation led to (what I believe is) the first complete enumeration and several new examples of a new class of polyhedral object.

In most cases, each unit of an origami modular corresponds to a feature of an underlying polyhedron—faces, edges, vertices, or combinations thereof. The most elegant designs lock together securely without glue by inserting folded tabs into pockets. Often each edge is represented by one origami unit, forming a skeleton of the polyhedron.

In 1995, a new twist (no pun intended) on this concept was devised by Thomas Hull, in a model "Five Intersecting Tetrahedra," (a.k.a. "FIT") based on a well-known arrangement of tetrahedra, realized using a unit developed by Francis Ow. In Hull's model, tetrahedra were built from edge units in the usual way, but five such tetrahedra were interwoven in an unusual and beautiful way.

Figure 1. Hull's FIT.

This model was, to my knowledge, the only origami model of intersecting polyhedra. Could there be more? That, of course, would depend on whether there were more intersecting polyhedra themselves. However, these polyhedra were intersecting in an unusual way: Although the faces intersected, the edges did not. This is what allowed their realization with an origami edge unit. Thus, I would need to search for intersecting polyhedra with nonintersecting edges. Since the Greek prefix "poly-" means "many" (as in "polyhedron" meaning "many faces"), I coined the term "polypolyhedra" to describe a structure composed of multiple polyhedra with nonintersecting edges. These would allow realization using simple, identical edge units analogous to the units used in Hull's FIT.

A precise mathematical definition of a polypolyhedron is *a compound of multiple linked polyhedral skeletons with uniform nonintersecting edges*. That is, it is composed of *multiple* polyhedra—there must be more than one, and all of the polyhedra must be identical. They must be topologically *linked* to one another, i.e., if they were made of string, we could not take them apart without cutting the string. The edges should be *nonintersecting*, meaning no edge touches any other edge other than at its endpoints. (The faces, of course, intersect.)

Further, the edges must be *uniform*. "Uniform" here has a special meaning: All the edges are alike in the sense that if we are examining a particular edge, we can rotate the model to bring any other edge into the position of the first edge, and the new edge and its surroundings will look the same as the origami edge. What about vertices? Since every edge must "look alike," it is tempting to assume that every vertex must also be uniform. But, in fact, there can be two classes of vertices, as long as every edge has exactly one of each vertex type. Thus, while the edges are uniform (also called 1-uniform), the vertices can be either 1-uniform (only one type of vertex) or 2-uniform (two types of vertices).

We can make a few observations immediately. One is that the vertices must be at least 2-valent, which means that at least 2 edges meet at every vertex. If any vertex is 1-valent (only one edge meets at a vertex), then the attached edge can't be part of a loop—but loops are the only way that the constituent polyhedra can be linked. Note that nowhere do we stipulate that the facial polygons of the constituent polyhedra must be planar. We will, in fact, find many polypolyhedra whose facial polygons are skew (nonplanar).

Now, the uniformity of the edges means that the polypolyhedron must have certain rotational symmetries. A polypolyhedron, therefore, can be classified according to its rotational symmetry.

If we arbitrarily pick a single edge of the polypolyhedron to be the "reference edge", then every other edge can be obtained from the reference edge by applying a rotational transformation. Therefore, there is a one-to-one correspondence between each of the edges of a polypolyhedron and some rotational transformation.

The rotational symmetries of a polypolyhedron must form a finite rotation group. There are five families of rotation groups in three-dimensions, called the cyclic, dihedral, tetrahedral, octahedral, and icosahedral groups. The rotation group of any polypolyhedron must fall into one of these five families, which we will briefly describe. Each symmetry group can be associated with a polygon or polyhedron and its rotational axes. (See Figure 2.)

The cyclic and dihedral rotation groups are made by the set of rotations and (for dihedral) flips that leave a polygon of n sides unchanged. There is a different group for every value of $n \leq 2$, but we do not need to consider any of these groups. This is because any edge will only make orbits that lie in a plane and thus merely produce polygons in a plane—hardly a polypolyhedron.

The tetrahedral, octahedral, and icosahedral groups are all based on rotations of their respective polyhedra, as shown in Figure 2. The axes of rotation can be through vertices, the midpoints of edges, or the center of faces.

For the tetrahedral, octahedral, and icosahedral groups, the axes of rotation fall naturally into three sets, associated with the faces, edges, and vertices of the

| Cyclic Group | Dihedral Group | Tetrahedral Group | Octahedral Group | Icosahedral Group |
| Order n | Order $2n$ | Order 12 | Order 24 | Order 60 |

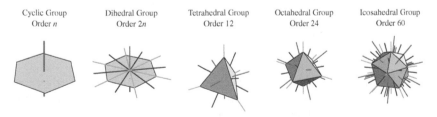

Figure 2. Illustrations of the symmetry groups in three-dimensions.

eponymous polyhedron. You might wonder: The rotation groups are named after three of the five Platonic solids. What about the other two solids, the hexahedron (cube), and dodecahedron? They are already taken care of: The former has octahedral symmetry; the latter has icosahedral symmetry. Every other polyhedron with rotational symmetry has symmetries that fall into one of these groups, and so does every polypolyhedron. In this work, I use a single-letter label for the rotation group: T, O, or I, for Tetrahedral, Octahedral, or Icosahedral symmetry.

Now, suppose we apply a rotational operator to a point. We get another point, which may or may not be distinct from the original point. The set of all distinct points obtained by applying all rotation operators in a rotation group to a single point is called an *orbit* of the point. The orbit is said to be *generated* by the initial (or *seed*) point and the rotation group. Sometimes, the number of points in an orbit—which is called the *order* of the orbit—is equal to the number of rotations in the group. However, any point that is on an axis of rotation is left unchanged by rotations about that axis. The orbits of such points are of lower order than the orbit of an arbitrary point. For example, for a tetrahedron (order = 12), the vertices of the tetrahedron form an orbit of order 4. The midpoints of the faces form another orbit of order 4. The midpoints of the edges form an orbit of order 6. Any point not lying on one of these lines is part of an orbit of order 12. Therefore, the tetrahedral group has several types of orbits with different numbers of points in the orbit. The same goes for the other rotational groups.

Let's give names to the three types of lower-order orbits. An orbit generated by a vertex of the eponymous polyhedron will be called a V orbit. An orbit generated by a midpoint of an edge will be called an E orbit; and an orbit generated by a midpoint of a face is an F orbit. Any other orbit will be called a C (for "complete") orbit. If a polypolyhedron has rotational symmetry, then any vertex of the polypolyhedron must lie in a V, E, F, or C orbit. The orders of the orbits of the various rotation groups are given in the following table.

	V	E	F	C
Tetrahedral	4	6	4	12
Octahedral	6	12	8	24
Icosahedral	12	30	20	60

Orbits matter because, if we select two vertices of any single edge, each vertex generates an orbit. The vertices of every other edge must lie within those two orbits (which may or may not be the same). If we pick a single edge (called the *seed edge*) of the polypolyhedron, one of its vertices defines an orbit;

every other edge must have one of its vertices in the same orbit. Similarly, the other vertex of the seed edge defines a second orbit; every other edge must have its other vertex in the same orbit. Therefore, we can further classify uniform-edge polypolyhedra by the orbits of their vertices in addition to their rotational symmetry group. In fact, we can fully characterize a uniform-edge polypolyhedron by its rotation group and a single edge, since all other edges may be obtained by application of the rotation operators of the symmetry group to that edge.

There are only four possible types of orbits for each vertex—V, E, F, or C. We label the seed edge with a symbol $\{o_1, o_2\}$, where $o_1, o_2 \in \{V, E, F, C\}$. We call this symbol the *orbital type* of the edge—and, by extension, of the polypolyhedron. For example, any edge of a $\{V,C\}$ uniform-edge polyhedron has one vertex in a V orbit and the other in a C orbit.

Note: We have 16 different orbital types.

In addition to these, there are two distinctly different types of double-letter combinations: A $\{V,V\}$ edge can have both vertices as part of the same V orbit or part of two different V orbits. (Two V orbits are different if the orbits lie at different radial distances from the origin.) We can distinguish between these cases by appending a number to the letter: $\{V1,V1\}$ denotes both vertices in the same orbit; $\{V1,V2\}$ means they lie in different orbits. Similarly, we have both $\{E1,E1\}$ and $\{E1,E2\}$, $\{F1,F1\}$ and $\{F1,F2\}$, and $\{C1,C1\}$ and $\{C2,C2\}$. So, in principle, there is a total of 20 different orbital types to consider. Not all 20 types are distinct, however. There is no distinction between a $\{V,C\}$ edge and a $\{C,V\}$ edge, other than the order in which we take the vertices for the seed edge. So we can eliminate some pairs as being redundant. After eliminating redundant types, we are left with 14 orbital types for each symmetry group: $\{V1,V1\}$, $\{V1,V2\}$, $\{V1,E1\}$, $\{V1,F1\}$, $\{V1,C1\}$, $\{E1,E1\}$, $\{E1,E2\}$, $\{E1,F1\}$, $\{E1,C1\}$, $\{F1,F1\}$, $\{F1,F2\}$, $\{F1,C1\}$, $\{C1,C1\}$, and $\{C1,C2\}$. Any polypolyhedron must be at least one of these types. (It is possible for a polypolyhedron to fall into more than one, since the V and F orbits of the tetrahedral group are equivalent to each other under a rotation.) But these are all there are, so we can enumerate all of the polypolyhedra of each orbital type, we can be assured that we have found all possible polypolyhedra.

Now that looks like there are a lot of polypolyhedra. For each rotation group T, O, and I, there are 14 possible pairs of orbits for the seed edge vertices, and within each orbital type, there is a large number of distinct edges (since we have a choice of many points within each orbit to use to construct the seed edge). However, we can eliminate some of the configurations as being uninteresting right off the bat by considering the valency of the vertices of a polypolyhedron. It is easily shown that all orbits of type $\{o,C\}$, where o is anything but C, yield univalent polypolyhedra (where the "polyhedra" are disjoint line segments). Thus, there are only 10 distinct types of edges that lead to interesting polypolyhedra, and let us denote these types by $\mathcal{T}$. So, $\mathcal{T} = \{\{V1,V1\}, \{V1,V2\}, \{V1,E1\}, \{V1,F1\}, \{E1,E1\}, \{E1,E2\}, \{E1,F1\}, \{F1,F1\}, \{F1,F2\}, \{C,C\}\}$.

We now define some more terms: a *homoörbital polypolyhedron* is one in which both vertices lie in the same orbit (the polyhedron is therefore 1-uniform-vertex); a *heteroörbital polypolyhedron* is one in which there are two distinct orbits (and the polyhedron is 2-uniform-vertex). In a heteroörbital polypolyhedron, every edge has one vertex in one orbit and one in the other orbit. The homoörbital polypolyhedra have edges of type $\{V1,V1\}$, $\{E1,E1\}$, $\{F1,F1\}$, and $\{C,C\}$; the heteroörbital polypolyhedra have edges of type $\{V1,V2\}$, $\{V1,E1\}$, $\{V1,F1\}$, $\{E1,E2\}$, $\{E1,F1\}$, and $\{F1,F2\}$.

Now, we can further classify polypolyhedra within a single orbit type. To collect every possible edge within an orbit of type $\{o_1, o_2\}$, we need only pick a single vertex from orbit o_1 as the first vertex of our seed edge, since every other vertex in the orbit is related by a simple rotation. For the second vertex, in principle, we should consider each of the $O(o_2)$ points in the o_2 orbit as the other vertex of the seed edge. If the first vertex is on an axis of rotation, however, there are rotations that leave the first vertex unchanged but not the second, meaning that this list of $O(o_2)$ seed edges contains some pairs that are equivalent under rotation. By considering all $O(o_2)$ possibilities and then eliminating duplicates (edges equivalent under rotation), we can be assured of finding every possible edge within the orbit type. So, by considering all ten orbit types for each symmetry group, and all five symmetry groups, we can identify every possible edge type and, therefore, every possible polypolyhedron.

Each polypolyhedron can be labeled by four quantities: its symmetry group, the orbit of the first vertex in the seed edge, the orbit of the second vertex, and an indicator of which point in the second orbit is the second vertex. We chose a labeling system based on the original rotation operators. For each symmetry group G, the rotation operators are labeled g_i, $i = 1, ..., N$, where N is the order of the group G. Define a point $p^{(G,o_2)}$ in orbit o_2 of group G; then the point $g_i \cdot p^{(G,o_2)}$ is also a point in the same orbit for every rotation operator g_i. Conversely, for any point q in the orbit o_2, we can find at least one operator g_i such that $q = g_i \cdot p^{(G,o_2)}$ for a reference point $p^{(G,o_2)}$. (There can be more than one such operator; for V, E, and F orbits, a given g_i, and therefore i, is not unique, for a given q.) We can therefore label the second vertex by the index i of the rotation operator that transforms the initial point $p^{(G,o_2)}$ into it. Thus, for symmetry group G having symmetry operators g_i, orbits o_1 and o_2, and second orbit index i, the two vertices of the seed edge will be given by $p^{(G,o_1)}$ and $g_i \cdot p^{(G,o_2)}$, respectively; and this polypolyhedron can be concisely denoted by the symbol $\{G \; o_1 \; o_2 \; i\}$.

So, finally, we have a method to enumerate all possible polypolyhedra. For each symmetry group $G \in \{T, O, I\}$ and orbit types $\{o_1, o_2\} \in \mathcal{T}$, we pick a point in the orbit $p^{(G,o_1)}$ to be the "seed point" for that orbit. Then the seed edge is the line segment from $p^{(G,o_1)}$ to $g_i \cdot p^{(G,o_2)}$ for all possible values of $i = 1, ..., O(o_2)$. This prescription gives some duplicates, since if o_2 is not a C orbit there are, by definition, multiple rotation operators g_i that give the same second vertex $g_i \cdot p^{(G,o_2)}$. The important thing is that every possible seed edge (and hence every possible polypolyhedron) will be given by this prescription. By constructing all possible seeds edges and sorting out the duplicates, We can be assured of finding every possible polypolyhedron.

And so, that is what I did, using *Mathematica*. I set up the operators for each of the rotation groups, constructed representative orbits, and then for each orbit type, constructed and sorted the unique edges within that orbit type. I constructed all possible edges of all possible groups and orbit types, obtaining 188 distinct edges—no more. Each edge is potentially a seed for a polypolyhedron. With each of those 188 edges, if we apply the rotation operators from the symmetry group, we obtain a set of edges that, taken together, potentially form a polypolyhedron. My *Mathematica* search found 188 combinations of the symmetry groups (T, O, I), the ten valid and distinct orbit combinations in $\mathcal{T}$, and the assorted values of second orbit indices, which gave distinct seed edges. Because of rotational symmetry, applying a rotation operator g_i should give another edge of the polypolyhedron; applying all rotation operators should, and does, give all of the edges. And so, we now have a complete list of all possible polypolyhedra.

Well, not quite. What I have done so far enforces rotational symmetry, uniform-edge, and 1- or 2-uniform vertices. However, I listed six conditions at the very beginning that were requirements for interesting polypolyhedra, and many of the 188 fail on one or more counts. Some of them are simple polyhedra, not compounds of multiple polyhedra. Some still have univalent vertices (I had eliminated combinations that were guaranteed to be univalent, but I had not done anything that absolutely insured multi-valency); some have intersections among the edges. All of these are no-no's as far as interesting polypolyhedra are concerned. After constructing all 188 polypolyhedra, I used *Mathematica* again to weed out univalent or intersection-containing polyhedra, which left 41 polypolyhedra behind. After computing the vertices and edges of the 41 polypolyhedra, I could also use *Mathematica* to compute some of the other properties, such as the number of polypolyhedra, the valencies of the edges, and so forth. I introduce here two notations: First, the edge valency is given as "N-M," where N and M are the vertex valencies of the seed edge ("2-3" means one vertex is 2-valent and the other is 3-valent). Second, the structure of the polypolyhedron will be described by a symbol $(P \times N \times M)$, where P

I Fl F2 7, 3- 3, 5×6×4   I C C 14, 2- 2, 12×1×5   I C C 12, 2- 2, 20×1×3

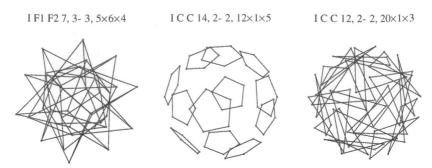

Figure 3. Skeletons for three of the 41 polypolyhedra.

is the number of disjoint polyhedra, N is the number of faces (polygons) per polyhedron, and M is the number of vertices (or edges) per polygon.

I then picked a seed edge with the prescribed orbits and plotted the skeletons of all 41 polypolyhedra thus obtained. The labels for the plots are the symmetry group, vertex orbits, and orbit index for the second vertex; the valency of each edge and the number of polyhedra, faces per polyhedra, and vertices (edges) per face. While all 41 were indeed composed of multiple polyhedra and were at least two-valent at every vertex, some of them did not consist of linked polyhedra. Figure 3 shows three of the skeletons, all from the icosahedral group I, composed, respectively, of five hexahedra, 12 pentagons, and 20 triangles.

But I cannot dismiss an unlinked polypolyhedron out of hand. Take, for example, the polypolyhedron {I C C 14} (middle). Clearly, the polyhedra (triangles) are unlinked. But that is only for the particular value of the seed point that I chose as the initial orbital seed. I could have chosen any other value of $p^{(I,C)}$ that did not fall on an axis of rotation; and it could be that one of the other values does indeed give a linked polypolyhedron (this actually turns out to be the case). Since I can vary the position of $p^{(I,C)}$ in two dimensions, there must be a continuous two-dimensional family of polypolyhedra associated with just the {I C C 14} polypolyhedron. In fact, there will be such a family for every {C,C} polypolyhedron.

The situation is a bit different for the heteroörbital polypolyhedra. The vertices lie in two different orbits, which correspond to two sets of axes of rotation of the polypolyhedron. But I can choose the radial distances of those two orbits independently; or if we neglect overall scale, I can vary the ratio of the orbital distances. Thus, there must be a continuous one-dimensional family of heteroörbital polypolyhedra.

So, I wasn't out of the woods yet. But the possibilities were down to a manageable number. The 41 skeletons represent all possible polypolyhedra. They fall naturally into three groups.

The first group is the homoörbital non-{C,C} polypolyhedra. There are only four of these, and apart from a constant scaling factor and mirror images, they are unique. The second group is the heteroörbital polypolyhedra. There are nine of these. But as I said above, each is really a family, characterized by a 1-dimensional parameter. The third group is the homoörbital {C,C} polypolyhedra. There are 28 of these. Each of these is characterized by a two-dimensional parameter set.

Now, to convert a polypolyhedral skeleton to an origami edge unit, imagine that each stick starts out infinitesimally thin, like a long, skinny balloon. We slowly inflate the balloon, inflating each stick at the same rate. As the sticks inflate, each point expands away from the center of the stick at the same rate, so the cross section of the stick is a circle. However, when two sticks touch, the balloons cannot continue to expand at the point of intersection; and as the balloons continue to inflate, the point where two balloons touch expands to form a flat plane, like the intersection of two soap bubbles. At that point, the circular cross section of the stick must develop a flat spot where it touches another stick. If we keep inflating the sticks, the circular parts keep expanding away from the axis of the stick, but the flat parts just get wider and wider. We enforce convexity of the stick by now allowing "hollows" to develop. As the expansion continues, the stick will eventually bump into another stick, and its cross section will develop another flat spot. As the inflation continues, the flat spots increase, multiply, and eventually merge, until a point is reached when there are no circular parts left. The stick can inflate no more; its cross section consists entirely of straight lines so that it has a polygonal cross section; and this polygon defines the largest cross section that any stick can have. The result is a convex polyhedron for each stick. Any subset of this polyhedron can be substituted for the stick without introducing an intersection with another stick. Thus, I can select two faces of the polyhedron to be the two facets of an origami edge unit analogous to the edge unit used in Hull's FIT.

A similar analysis creates origami modulars of the four homoörbital polypolyhedra, shown in Figure 4.

The first three are compounds of "polyhedra" in which each polyhedron consists of a single polygon (respectively, four interesecting triangles, six pentagons, and ten triangles). The last is equivalent to Hull's FIT. Instructions for folding the $(4 \times 1 \times 3)$ (first on left) and the $(10 \times 1 \times 3)$ (third from left) may be found in [2].

In the heteroörbital polypolyhedra, the vertices come in two distinct orbits. Every vertex within a single orbit lies at the same orbital distance, but the orbital distances can vary completely independently. Any given vertex has its location defined by ρ, θ, ϕ in polar coordinates. The angles θ and ϕ are fixed by the orbit—the point must lie on an axis of rotation—but the radial coordinate

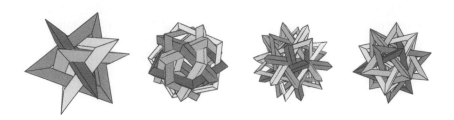

Figure 4. The four homoörbital polopolyhedra in origami.

ρ can vary. Since we have two sets of vertices, we have two radial parameters, ρ_1 and ρ_2, which can be chosen freely. However, if we further stipulate that the polypolyhedron has an edge of unit length, then we no longer can choose the radial parameters independently; we can only choose their ratio ρ_1/ρ_2. We can parameterize the two radial distances on a single parameter x that varies from 0 to 1. We define a normalizing function $f(x)$ such that $\rho_1 \equiv f(x)x$, $\rho_2 \equiv f(x)(1-x)$, and the normalizing function $f(x)$ is chosen to ensure unit edge length for all values of x. Then, as we sweep x from 0 to 1, we will trace out all possible values of the ratio ρ_1/ρ_2 from 0 to ∞.

Now, we introduce a function $d_{\min}(x)$, which is defined as the minimum distance between the seed edge and any other edge in the polypolyhedron. Since (by definition) the edges of polypolyhedra do not intersect, if we plot this function, it will be nonzero for a valid polypolyhedron, which is the case for most values of x. However, at some value(s) of x, it drops to zero. The zero values occur when the seed edge intersects one or more other edges of the polypolyhedron. As x moves through one of these zero values, the linkage between a polyhedron containing the seed edge and a polyhedron containing the intersecting edge changes. This change means that the polypolyhedra obtained for x values on either side of a zero of $d_{\min}(x)$ are topologically distinct. Although they have the same number of polyhedra, faces, vertices, and vertex valencies, the polyhedra are linked in different ways.

Now, we can divide the nine heteroörbital polypolyhedra into two groups: those in which the two orbits are of the same basic type, but different radial distances (e.g., {E1,E2} or {F1,F2}), and those in which the two orbits are entirely different (e.g., {E1, F1}). We call the formers—where the two types of vertices lie in the same orbit, but at different orbital distances—*quasihomoörbital* polypolyhedra. Quasihomoörbital polypolyhedra have a basic symmetry with respect to the variable x. The polypolyhedron obtained by making the substitution $x \rightarrow (1-x)$ is the same polypolyhedron, but the roles of E1 and E2 (or F1 and F2) are reversed.

I E1 E2 6, 2- 2, 6×1×10

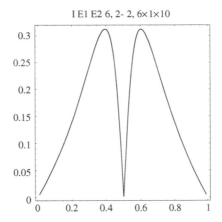

Figure 5. Example of a graph of $d_{\min}(x)$.

Thus, if we plot $d_{\min}(x)$ for the five quasihomoörbital polypolyhedra, we indeed find that $d_{\min}(x)$ is symmetric about $x = 0.5$, where there is a zero, as in the example in Figure 5. Because of this symmetry, the values x and $(1-x)$ give the topologically same polypolyhedron; thus, there is one polypolyhedron for each of the five symmetry group/orbit/orbital seed combinations.

The other four heteroörbital types are, however, truly heteroörbital: The vertices lie in two entirely different types of orbits. A plot of $d_{\min}(x)$ does not show a symmetry around $x = 0.5$. In a true heteroörbital polypolyhedron, there is no more symmetry with respect to the parameter x, so each lobe of each plot defines a topologically distinct polypolyhedron. So, for example, the {I E1 F2 7} polypolyhedron has three topologically distinct variants. In total, there are two types of {O E1 F1 4}, four of {I V1 E1 4}, two of {I E1 F1 6}, and three of {I E1 F1 7}, for a total of 11 topologically distinct polypolyhedra. Add these to the nine we found so far, and we're up to 20.

Now, things start to get complicated. I'd found 28 {C,C} orbital polypolyhedra, but those 28 were found for a single orbital seed point $p^{(G,C)}$. But that point can vary in two dimensions. Every possible value in the two-dimensional continuum of $p^{(G,C)}$ can be mapped to each of the 28 {C,C} candidates.

However, there is some duplication among these 28. All 28 can be classified into one of six groups according to their group, edge valencies, and the number and type of intersecting polyhedra (see Table 1).

Setting aside the dihedral group for a moment, consider the four tetrahedral types. Each is composed of four triangles ($4 \times 1 \times 3$), and the edges have tetrahedral rotation symmetry, which means that they are very similar. In fact, each of the four can be obtained from any of the others by rotating the four

Group	Edge valency	Number of polyhedra	Faces per polyhedron	Edges per face
Tetrahedral	2-2	4	1	3
Octahedral	2-2	6	1	4
Octahedral	2-2	8	1	3
Icosahedral	2-2	12	1	5
Icosahedral	2-2	20	1	3

Table 1.

triangles about their centers and/or translating the triangles with respect to the origin. This is a general transformation however: Any possible arrangement of four triangles with tetrahedral rotational symmetry should be obtainable by the same transformation. So, we can drop all but one of the $(4 \times 1 \times 3)$ $\{C,C\}$ configurations and look at how any one of them varies under this facial rotation/translation transformation.

Imagine rotating each triangle in a tetrahedron by an angle ϕ and translating it radially so that its centroid lies at a distance ρ from the origin. For every $\{\phi, \rho\}$ pair, we obtain a unique arrangement of triangles—and for some $\{\phi, \rho\}$, we will obtain every possible arrangement of equilateral triangles with tetrahedral symmetry. Let's also call up our old friend $d_{\min}$, the shortest distance between the reference edge and any other edge. In the heteroörbital $\{C, C\}$ polypolyhedra, $d_{\min}$ must vary with two parameters—ϕ and ρ. So, $d_{\min}$ is now a two-dimensional plot. In Figure 6, I plotted $d_{\min}(\phi, \rho)$ as an intensity plot,

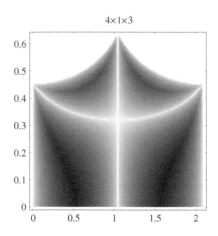

Figure 6. Example of the intensity plot of $d_{\min}(\phi, \rho)$.

where darker areas correspond to larger values of the function. The horizontal axis is rotation angle (ϕ) and the vertical axis is distance from the center (ρ). Darker regions thus correspond to greater distances between the seed edge and any other edge. Obviously, the function is periodic in ϕ with period 2π (imagine the plot wrapped around like a vertical cylinder). As with the heteroörbital polypolyhedra, crossing a zero of $d_{min}(\phi, \rho)$ means the topological linkage has changed; thus, each shaded region corresponds to a topologically distinct polypolyhedron. It is apparent from the plot that there are four such regions (the right and left edge of the plot should be thought of as joined). Two of them are mirror images of the other two. (Mirror image polypolyhedra lie symmetrically about the obvious vertical symmetry lines of the plot.) So, not counting mirror images, there are two distinct tetrahedral {C,C} polypolyhedra.

The one at $\{\phi \approx 1.3, \rho \approx 0\}$ turns out to be topologically identical to the $(4 \times 1 \times 3)$ that we already found, the homoörbital {O E1 E2 2}. (The difference is that, in the former, the planes of the polygons all intersect the origin; in the tetrahedral {C,C} polypolyhedron, the four triangles are offset from the origin). But there is another tetrahedral polypolyhedron, located at $\{\phi \approx 1.03, \rho \approx .203\}$, which constitutes a different arrangement of four triangles. Thus, there are two topologically distinct heteroörbital {C,C} polypolyhedra.

Next come the two possibilities for octahedral {C,C} orbits. We can let {O C C 4} stand in for all possible $(6 \times 1 \times 4)$ polypolyhedra. There turn out to be two distinct octahedral $(6 \times 1 \times 4)$ polypolyhedra. The other octahedral $\{C, C\}$ type is the $(8 \times 1 \times 3)$, which has three topologically distinct arrangements. In the icosahedral group, the $(12 \times 1 \times 5)$ family yields five topologically distinct arrangements plus their mirror images. Last, the $(20 \times 1 \times 3)$ set, which leads to an astonishing total of 23 different varieties (plus mirror images, of course).

When we put together all of the polypolyhedra of the various types, we find a total of exactly 54 topologically distinct polypolyhedra, not counting rotations and reflections. Nearly half of them are $(20 \times 1 \times 3)$. All of them are excellent candidates for origami implementation. The structure and relationships among the various polypolyhedra are summarized in Table 2. Out of all 54 polypolyhedra, all but two have at least some 2-valent vertices. The two exceptions are the $(5 \times 4 \times 3)$ and a quasihomoörbital polypolyhedra—the {I F1 F2 7} polypolyhedron, whose structure is $(5 \times 6 \times 4)$. This property sets these two apart from the others. The $(5 \times 4 \times 3)$, of course, is the basis of Hull's FIT, but the $(5 \times 6 \times 4)$ had not been implemented in origami. It, too, possesses a simple 60-unit implementation and is shown in Figure 7. Folding instructions may be found in [2]

Now, final notes on some prior art. In the origami world, Hull's Five Intersecting Tetrahedra has been gathering *oohs* for several years. Outside origami, a number of the polypolyhedral structures have shown up singly in

Figure 7. The $(5 \times 6 \times 4)$.

the geometrical arts. Perhaps the most extensive collection is to be found in *Orderly Tangles*, by Alan Holden [3], in which the author shows examples of about half of the polypolyhedra shown here built as models from wooden dowels (which he calls "polylinks"). However, to my knowledge, no one has enumerated all of the polypolyhedra; this work is the first complete enumeration of polypolyhedra and regular polylinks. So what started out as a fairly simple question turned out to lead to some pretty amazing—and perhaps even new—mathematics (and origami, too).

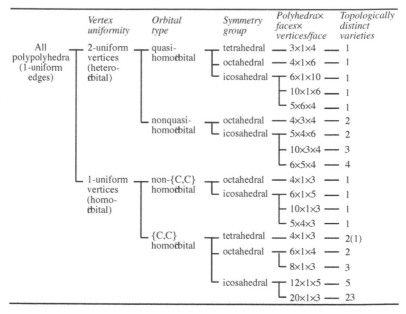

	Vertex uniformity	Orbital type	Symmetry group	Polyhedra× faces× vertices/face	Topologically distinct varieties
All polypolyhedra (1-uniform edges)	2-uniform vertices (hetero-orbital)	quasi-homorbital	tetrahedral	3×1×4	1
			octahedral	4×1×6	1
			icosahedral	6×1×10	1
				10×1×6	1
				5×6×4	1
		nonquasi-homorbital	octahedral	4×3×4	2
			icosahedral	5×4×6	2
				10×3×4	3
				6×5×4	4
	1-uniform vertices (homo-orbital)	non-{C,C} homorbital	octahedral	4×1×3	1
			icosahedral	6×1×5	1
				10×1×3	1
				5×4×3	1
		{C,C} homorbital	tetrahedral	4×1×3	2(1)
			octahedral	6×1×4	2
				8×1×3	3
			icosahedral	12×1×5	5
				20×1×3	23

Table 2. Polypolyhedra taxonomy.

Acknowledgements

I would like to acknowledge helpful commentary and suggestions from Thomas Hull and sarah-marie belcastro, Stan Isaacs (who put me onto Holden's book), John H. Conway (whose analysis of 100-hedra in a mathlist posting suggested the key approach), and Péter Budai, who actually folded the $(5 \times 6 \times 4)$.

References

[1]   Hull, T. "Five Intersecting Tetrahedra", appears in *Origami: The Complete Guide to the Art of Paperfolding* (Beech, R., Lorenz Books, 2001). See also http://web.merrimack.edu/~thull/fit.html.

[2]   *2000 Annual Convention*, New York: OrigamiUSA, (2000).

[3]   Holden, Alan, *Orderly Tangles: Cloverleafs, Gordia Knots, and Regular Polylinks*, (New York: Columbia University Press, 1983).

Origami with Trigonometric
Functions

Miyuki Kawamura

1 Do You Love Trigonometric Functions?

In Japan, we learn trigonometric functions in the high school math program. These are very beautiful functions, in my opinion, but many people hate them. Why? Some people say, "Because these are useless in my life", or "They are difficult". I believe that trigonometric functions are very useful tools and they should be more familiar to origami artists. These functions express the relation between length and angle. Length is easy to handle via paper-folding because we know various origamic ways to divide an edge of paper. On the other hand, almost all angles are hard to handle except for specific ones, for example 90, 60, 30, and 45 degrees.

2 What are Trigonometric Functions?

The trigonometric functions are given in Figure 1.

$$\sin\theta = \frac{c}{a}, \quad \cos\theta = \frac{b}{a}, \quad \tan\theta = \frac{c}{b}$$

These equations show us the relations between the angle θ and the ratios of the lengths of sides a, b and c on a right triangle. This is a reason why these functions are useful, because we know many ways to divide a length into any

169

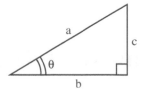

Figure 1. The definition of sine, cosine, and tangent.

ratios by folding. So, we can get some lengths on a paper easily and make some right triangles, too. When the angle of θ is greater than 90 degrees, the trigonometric functions are given by

$$\sin \theta = \frac{c}{a}, \quad \cos \theta = \frac{b}{a}, \quad \tan \theta = \frac{c}{b} \quad (0° \leq \theta \leq 90°) \tag{1}$$

and

$$\sin \theta = \frac{c}{a}, \quad \cos \theta = -\frac{b}{a}, \quad \tan \theta = -\frac{c}{b} \quad (90° \leq \theta \leq 180°). \tag{2}$$

3   How to Make Angles with Trigonometric Functions

I will show three different ways to make an angle of θ on a paper by folding with trigonometric functions. The arbitrary angle θ satisfies Equations (1) and

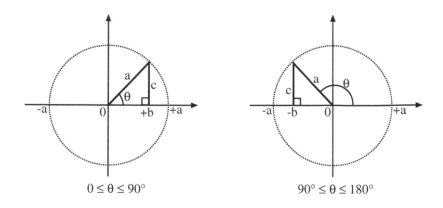

Figure 2. The difference between acute and obtuse angles.

(2). Here $a, b,$ and c, which are some real numbers, denote the lengths of sides on some right triangle (Figure 1) and $0 \leq b, c \leq a$.

3.1 How to Use the Cosine Function

Here is a way to make an angle of θ with the cosine function. The angle θ satisfies $\cos \theta = \frac{b}{a}$. where $0° \leq \theta \leq 90°$ and $0 \leq b \leq a$.
  You can do the following steps (see Figure 3):

1. Decide the distance "a".

2. Decide the distance "b".

3. Put the endpoint of line "a" on the line "b" and fold.

4. Do the same on the other side and you are finished.

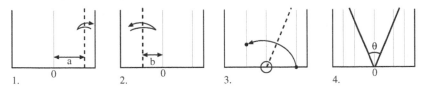

Figure 3. Using cosine to construct an arbitrary angle θ.

  The reason why you end up with the angle θ is as follows (see Figure 4): The angle A'0B is θ because the value of the cosine function of the angle A'0B is b/a. So the angle F0A is $(180° - \theta)/2$ and then the angle 0'0F is $\theta/2$.

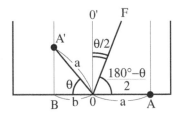

Figure 4. Proof of the previous method.

  When the angle of θ is greater than 90 degrees, θ satisfies the equation $\cos \theta = -\frac{b}{a}$, where $90° \leq \theta \leq 180°$ and $0 \leq b \leq a$. Then you can do the following steps (see Figure 5):

1. Decide the distance "a".

2. Decide the distance "$-b$".

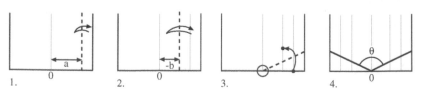

Figure 5. Using cosine when θ is obtuse.

3. Put the endpoint of line "a" on the line "$-b$" and fold.

4. Do the same on the other side and you are finished.

(This method with the cosine function is not so practical when the angle θ is close to 0 or 180 degrees, because an infinitesimal error in folding makes a large error on the angle θ.) For example, when the angle θ is 60 degrees, the value of the cosine function is $\cos 60° = 1/2$ and you can do the following steps (see Figure 6).

1. Make a line on the center of the paper.

2. Fold the quarter line.

3. Put the corner on the quarter line and fold.

4. Do the same on the other side and you are finished.

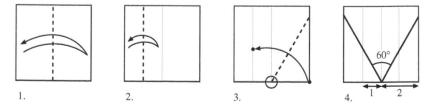

Figure 6. Using cosine to construct a 60° angle.

These are the same steps that we usually use to make 60 degrees.

3.2   How to Use the Sine Function

This is a way to make an angle of θ with the sine function. The angle θ satisfies the equation $\sin \theta = \frac{c}{a}$, where $0° \leq \theta \leq 90°$ and $0 \leq c \leq a$.

You can do the following steps (see Figure 7).

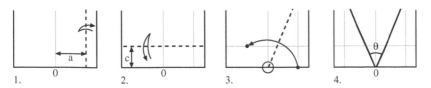

Figure 7. Using sine to construct an arbitrary angle θ.

1. Decide the distance "a".

2. Decide the distance "c". This is not the direction for cos.

3. Put the endpoint of line "a" on the line "c" and fold.

4. Do the same on the other side and you are finished.

The reason why this makes an angle θ is as follows (see Figure 8): The angle A′0B is θ because the value of the sine function of the angle A′0B is c/a. So the angle F0A is $(180° - \theta)/2$ and then the angle 0′0F is $\theta/2$.

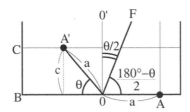

Figure 8. Proof of this method.

When the angle of θ is greater than 90 degrees, θ satisfies $\sin\theta = \frac{c}{a}$, where $90° \leq \theta \leq 180°$ and $0 \leq c \leq a$, and you can do the following steps (see Figure 9).

1. Decide the distance "a".

2. Decide the distance "c".

3. Put the end point of line "a" on the line "c" and fold.
   Note: This could have been done in Figure 7 if we were not careful.

4. Do the same on the other side and you are finished.

(This method with the sine function is not so practical when the angle θ is close to 90 degrees, because of the same reason as cosine.) For example, when the angle θ is 30 degrees, the value of the sine function is $\sin 30° = 1/2$ and you can do the following steps (see Figure 10).

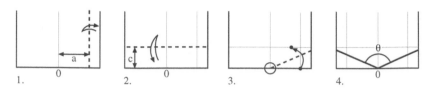

Figure 9. Using sine when θ is obtuse.

1. Make two lines on the center of a paper.
2. Make the quarter line.
3. Put the corner on the quarter line and fold.
4. Do the same on the other side and you are finished.

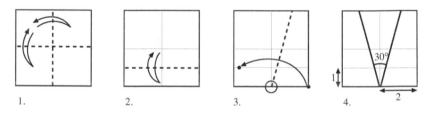

Figure 10. Using sine to construct 30°.

3.3 How to Use the Tangent Function

This is a way to make an angle of θ with the tangent function. The angle θ satisfies the equation $\tan \theta = \frac{c}{b}$, where $0° \leq \theta \leq 90°$ and $0 \leq b, c$.

You can do the following steps (see Figure 11).

1. Decide the distance "c".
2. Decide the distance "b".

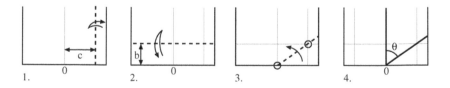

Figure 11. Using tangent to construct an arbitrary angle θ.

3. Fold on the line which passes through the intersection.

4. You are finished.

The reason why we get the angle θ is trivial by Figure 11 Step 3. When the angle of θ is greater than 90 degrees, θ satisfies $\tan\theta = \frac{c}{-b}$, where $90° \le \theta \le 180°$ and $0 \le b, c$. and you can do the following steps (see Figure 12).

1. Decide the distance "c".

2. Decide the distance "$-b$".

3. Fold on the line which passes through the intersection.

4. You are finished.

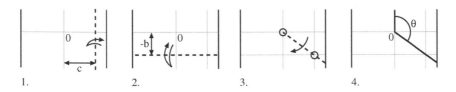

1.   2.   3.   4.

Figure 12. Using tangent when θ is obtuse.

(In this method, the tangent function is not so practical when the angle θ is close to 0 or 180 degrees, for the same reason as cosine.) For example, when the angle θ is 45 degrees, the value of the tangent function is $\tan 45° = 1/1$ and you can do the steps in Figure 13. (This is the usual way to make an angle of 45 degrees, of course!)

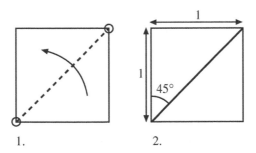

1.   2.

Figure 13. We've been using tangent all along to construct 45°.

4 How to Make Angles Using Approximations

The three above examples are very simple, but in general, the values of a, b, and c will be more complicated. For example,

$$\cos 36° = \frac{1 + \sqrt{5}}{4}.$$

We can make exactly the angle of 36 degrees using this value, of course. But this is a little complicated when all you want to do is make some ornaments for your Christmas tree! So we will approximate the value by a ratio of 2 integers. The angle of 36 degrees satisfies the following equation;

$$\cos 36° = \frac{1 + \sqrt{5}}{4} = 0.809016994 \cdots$$

and

$$\cos 36° = n \times \frac{\cos 36°}{n} = n \times \frac{0.809016994 \cdots}{n}.$$

Then, if

$$n = 1 \longrightarrow \cos 36° = \frac{0.8090 \cdots}{1},$$

$$n = 2 \longrightarrow \qquad = \frac{1.6180 \cdots}{2},$$

$$n = 3 \longrightarrow \qquad = \frac{2.4270 \cdots}{3},$$

$$n = 4 \longrightarrow \qquad = \frac{3.2360 \cdots}{4},$$

$$n = 5 \longrightarrow \qquad = \frac{4.0450 \cdots}{5} \approx \frac{4}{5} \quad \text{etc.}$$

If we increase n, eventually the fraction will be close to an integer fraction. This gives us the approximate values

$$\cos 36° \approx \frac{4}{5} \quad \text{or} \quad \frac{13}{16} \quad \text{or} \quad \frac{17}{21} \quad \text{or} \quad \cdots.$$

We can then make approximately the angle of 36 degrees by the method shown in Section 3.1. The extention to other angles is also straightforward.

5 Tables

It is convenient to have a table of the values of the trigonometric functions for some specific angles. Here is a special table that I use in my origami works. You can make special tables for your models, too.

Archmedean Dual Polyhedron		θ (degrees)	sinθ (error:%)	cosθ (error:%)	tan(θ/2)(error:%)
Triakis Tetrahedron		112.8853805...	11/12 (+0.59)	-2/5 (+0.61)	3/2 (-0.24)
		33.5573098...	5/9 (+0.57)	5/6 (0)	1/3 (+9.87)
Triakis Octahedron		117.2005704...	8/9 (+0.06)	-5/11 (-0.14)	5/3 (+0.74)
		31.3997148...	1/2 (-4.46)	6/7 (-1.26)	2/7 (+1.56)
Tetrakis Hexahedron		83.6206298...	1/1 (+7.63)	1/9 (0)	8/9 (-0.42)
		48.1896851...	3/4 (+0.83)	2/3 (0)	4/9 (-0.55)
Triakis Icosahedron		119.0393509...	7/8 (-0.07)	-1/2 (+0.81)	5/3 (-0.81)
		30.4803246...	1/2 (-1.58)	6/7 (+1.71)	1/4 (-7.90)
Pentakis Dodecahedron		68.6187209...	14/15 (+0.50)	1/3 (+2.78)	2/3 (-1.81)
		55.6906395...	5/6 (+1.35)	4/7 (-0.97)	1/2 (-4.60)
Rhombic Dodecahedron		109.4712206...	15/16 (+0.82)	-1/3 (0)	7/5 (-0.50)
		70.5287794...	15/16 (-1.27)	1/3 (0)	5/7 (+0.77)
Rhombic Triacontahedron		116.5650512...	8/9 (+0.60)	-4/9 (-0.15)	8/5 (-0.49)
		63.4349488...	8/9 (-1.11)	4/9 (+0.28)	3/5 (-2.38)
Trapezoidal Icositetrahedron		115.2631744...	9/10 (+0.50)	-3/7 (+0.10)	11/7 (-0.18)
		81.5789419...	1/1 (+10.3)	1/7 (+0.25)	6/7 (-0.46)
Trapezoidal hexecontahedron		118.2686775...	7/8 (+0.58)	-1/2 (+1.46)	5/3 (-0.17)
		86.9741555...	1/1 (+3.48)	1/19 (+0.01)	1/1 (+3.48)
		67.7830115...	13/14 (+0.63)	3/8 (+0.28)	2/3 (-0.59)
Hexakis Octahedron		55.0246961...	4/5 (-3.44)	4/7 (+0.23)	1/2 (-3.44)
		87.2019638...	1/1 (+3.21)	1/21 (+0.08)	1/1 (+3.21)
		37.7733401...	3/5 (-2.39)	4/5 (-2.39)	1/3 (-2.39)
Hexakis Icosahedron		58.2379196...	6/7 (+1.30)	1/2 (+3.03)	5/9 (-0.22)
		88.9918019...	1/1 (+1.13)	0 (+1.13)	1/1 (+1.13)
		32.7702785...	1/2 (-8.45)	5/6 (+2.40)	2/7 (-2.68)
Pentagonal Icositetrahedron		80.7517020...	1/1 (+11.5)	1/6 (-0.43)	6/7 (+0.56)
		114.8120744...	10/11 (-0.17)	-2/5 (-1.07)	11/7 (+0.21)
Pentagonal Hexehedron		67.4535092...	12/13 (-0.12)	2/5 (-1.53)	2/3 (-0.11)
		118.1366227...	8/9 (-0.74)	-1/2 (+1.58)	5/3 (-0.05)

Table 1. Approximate integer fractions for making angles in the Archimedian dual polyhedra.

Regular Polygon	θ (degrees)	sinθ	(error:%)	cosθ	(error:%)	tanθ	(error:%)
	● = 60	6/7	(-1.67)	1/2	(0)	7/4	(+0.43)
	▲ = 120	6/7	(+0.84)	-1/2	(0)	-7/4	(-0.21)
	◆ = 30	1/2	(0)	6/7	(+3.34)	4/7	(-0.85)
	● = 90	1/1	(0)	0	(0)	∞	(-)
	◆ = 45	5/7	(+1.30)	5/7	(-1.30)	1/1	(0)
	● = 108	19/20	(+0.18)	-1/3	(+1.36)	-3/1	(+0.40)
	▲ = 72	19/20	(-0.27)	1/3	(-2.04)	3/1	(-0.60)
	◆ = 54	4/5	(-1.61)	3/5	(-1.61)	11/8	(-0.05)
	★ = 36	3/5	(+2.42)	4/5	(+2.42)	3/4	(+2.42)
	● ≈ 128.5714...	4/5	(-1.32)	-5/8	(+0.09)	-5/4	(+0.07)
	▲ ≈ 51.42857...	4/5	(+3.31)	5/8	(-0.22)	5/4	(-0.17)
	◆ ≈ 64.28571...	9/10	(-0.20)	3/7	(+0.52)	2/1	(-1.32)
	★ ≈ 25.71428...	3/7	(-1.31)	9/10	(+0.50)	1/2	(+3.31)
	● = 135	5/7	(-0.43)	-5/7	(+0.43)	-1/1	(0)
	▲ = 45	5/7	(+1.30)	5/7	(-1.30)	1/1	(0)
	◆ = 67.5	12/13	(-0.18)	2/5	(-1.60)	12/5	(-0.18)
	★ = 22.5	2/5	(+4.79)	12/13	(+0.53)	2/5	(-3.10)
	● = 140	2/3	(-1.29)	-3/4	(-1.01)	-5/6	(+0.14)
	▲ = 40	2/3	(+4.53)	3/4	(+3.52)	5/6	(-0.49)
	◆ = 70	15/16	(-0.52)	1/3	(+0.76)	11/4	(+0.02)
	★ = 20	1/3	(-2.64)	15/16	(+1.82)	1/3	(-7.83)
	● = 144	3/5	(-0.60)	-4/5	(-0.60)	-3/4	(-0.60)
	▲ = 36	3/5	(+2.42)	4/5	(+2.42)	3/4	(+2.42)
	◆ = 72	19/20	(-0.27)	1/3	(-2.04)	3/1	(-0.60)
	★ = 18	1/3	(+8.17)	19/20	(+1.08)	1/3	(+2.42)

Table 2. Approximate integer fractions for angles in regular polygons.

6   Conclusion

Trigonometric functions are very powerful tools in the origami world. These functions and origami are very good friends because these functions connect length with angle. The methods presented in this paper are very general, are very easy, and easy is good! I hope that the trigonometric functions become standard tools for origami-lovers.

References

[1] Kawamura, M., "Polyhedron origami: a possible formulation by 'simple units'", *Proceedings of The 2nd International Meeting of Origami Science and Scientific Origami*, Seian University of Art and Design, Otsu, Japan, (1995).

A Study of Twist Boxes

Noriko Nagata

1   Introduction

Figure 1 shows several twist boxes made of a single piece of square paper. At a glance, these boxes look like different types of boxes. However, as is seen in the Figure 1, they are based on a common diagram. Each diagram shows a square sheet of paper folded along the four edges at a specified width. Consider the inner square of each diagram. The four equivalent on each side of the inner square form a square. This becomes the bottom square of the box. Hereafter we will label one of these four points as point P. The triangle part above the point P is used for making a solid body by a common procedure shown in [1]–[5].

The different boxes can be characterized by the position of the point P and by the fold width. Naturally, when the fold width becomes larger, the bottom square becomes smaller. Thus when the point P is at the center of each side, one can make a cube. If we rotate the bottom square relative to the other folds, a twist box can be formed. If another point P is taken, a different type of twist box with different height and twist angle is constructed. So various twist boxes can be formed by the present procedure. In a special case, the boxes can be completely folded flat.

The purpose of this study is to clarify the relationship between the bottom square position on the original sheet and the completed box.

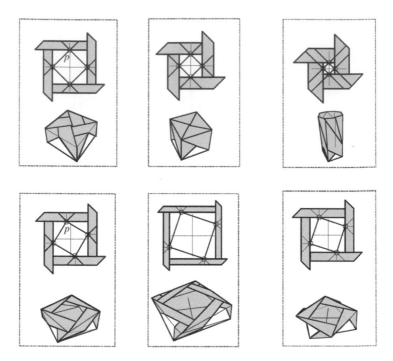

Figure 1. Twist boxes.

2   Two Types of Twist Boxes

There are two types of twist boxes, shown in Figure 2. There are four parallelograms in each diagram. They turn into the four sides of the completed box. When a mountain crease is made along the shorter diagonal of each parallelogram, as in Type A, a mountain fold twist box is made. In the same way, when we make a valley crease along the longer diagonal of each parallelogram, as in Type B, a valley fold twist box is formed.

Before several properties of the twist boxes can be discussed, the positions of the bottom square on the original sheet have to be specified. Let the length of a side of the original square be 2, and consider a quarter of the square (see Figure 3). Point P can be specified by two parameters, a and b. Here a is the distance from the center line and b is the fold width. So a can range between zero and $1 - 2b$ and the fold width b is limited to being less than $1/2$. Also, let θ be the twist angle, which is the angle difference between the (stationary) bottom square and the (twisted) upper square of the box.

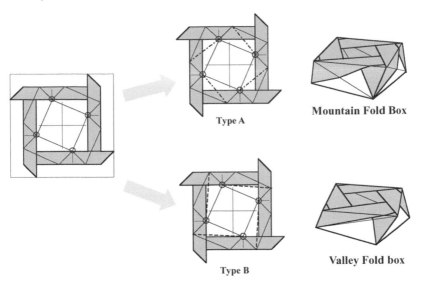

Figure 2. Two types of twist boxes.

For example, we start from the diagram for fold width $b = 1/6$ (Figure 4). In the case of a mountain fold (Type A of Figure 2), when point P moves from O to M, the height h of the box decreases monotonically, as shown in the curve in graph 1 (Figure 4). Thus point P can move in the whole range between O and M. On the other hand, in the valley fold case, as point P moves away from O, the height decreases as shown in the lower curve in graph 1 (Figure 4) and becomes zero at the point N lying somewhere short of M. So point P can move in the limited range from O to N and the value of a at point N is given by $1 - 3b$, as can be seen from a simple geometric argument. Graph 2 (Figure 4) shows the twist angle θ of the two types of twist boxes. When point

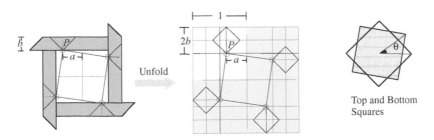

Figure 3. Specification of point P on the origami sheet.

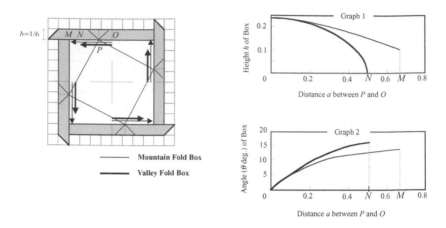

Figure 4. Trajectories of point P, height h, and angle of the box.

P moves away from O, the twist angles increase monotonically and reach a maximum value for both the mountain and valley fold box.

Here it should be notated that when point P coincides with point N, the box can be folded flat perfectly with the maximum twist angle.

3   The Mountain Fold Boxes

The curves in the left graph of Figure 5 are the calculated heights of mountain fold boxes for several values of the fold width b. The horizontal axis shows the distance a between P and O. As can be easily seen, each curve decreases

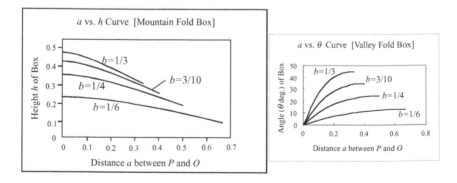

Figure 5. Height and twist angle of mountain folded boxes.

monotonically with increasing a and the range of point P is limited by $1 - 2b$. The curves in the right graph (Figure 5) are the calculated twist angles of mountain fold boxes. Every curves start from zero and increases monotonically with increasing a.

4   Valley Fold Boxes

The lines in the left graph of Figure 6 are the calculated heights of valley fold boxes. Contrary to the mountain fold ones, for valley fold boxes there exists, for every value of fold width b, the point P at which the height of box becomes zero and the box can be folded. Furthermore, for the rather small fold widths b, the height h decreases monotonically with increasing a. However, for b larger than 0.27, the a vs. h curves show some complicated behaviors. That is, there exist two boxes with different heights constructed for each point P in some range beyond point N at which the box can be folded. Therefore, 0.27 is a critical value of b for which the height h can be varied freely in a region between 0 and a finite value, and the twist angle at N is just 45 degrees (see the right graph in Figure 6).

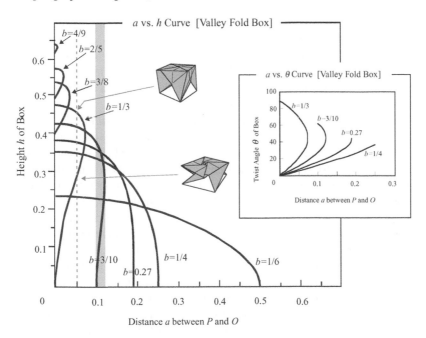

Figure 6. Height and twist angle of valley fold boxes.

Now we shall consider the curve for $b = 3/10$. As easily seen from this curve, when a is in the gray range, two different boxes can be constructed for each point P. In these cases, it should be noticed that the sum of the twist angles of the two boxes is 90 degrees. Increasing b slightly, we see that the curve of $b = 1/3$ is the special case where the grayed region starts at $a = 0$.

In this case, a cube box can be constructed at the point O and the height of a cube is given by the value of the top of the curve. As is well known, a cube box can be folded perfectly flat when it is twisted by 90 degrees (as indicated in the right graph of Figure 6). The pictures in the left graph of Figure 6 show two boxes at the same point P for $b = 1/3$. Of course, the sum of these twist angles is 90 degrees. On the other hand, for b beyond $1/3$ the curves show a different behavior. The important features of these curves can be given by the following three remarks. Firstly, the range of point P, at which a box can be constructed, is restricted to nearby point O. Secondly, two different boxes can always be constructed at each point P in the range. Third, these boxes can never be folded flat.

5   Diagram of the Point P

As mentioned above, if point P is taken anywhere (between O and M) then a mountain fold box having point P as a corner of the bottom square can always be constructed. However, the valley fold boxes can only be constructed for P in some restricted regions on the sheet. Therefore the valley fold boxes can be changed into the mountain fold ones, while the mountain fold boxes cannot always be changed into the valley fold ones.

Figure 7 shows a quarter of a sheet. Here the line m shows the trajectory of the point M which is given by $a = 1 - 2b$. The shadowed triangle ACE, which is 1/8 of the sheet, shows the regions of point P where we can construct mountain fold twist boxes. On the other hand, line n shows the trajectory of point N which is given by $a = 1 - 3b$. This line corresponds to the diagonal of the rectangle with ratio, 3 to 2. The boxes made from point P on this line are always folded entirely flat except for b beyond the critical value 0.27. Accordingly, the triangle ADE is the region where a single valley fold twist box can be constructed at each point P. Now, as indicated in the previous graph (Figure 6), for fold widths $b > 0.27$ two valley fold boxes with the different heights and twist angles can be made at each point P in some narrow range of a beyond the line n. This region is shown by the black bill-like zone in the Figure 7. These results can also be applied to the triangle ABC with 1/8 the area. However, the twist of the boxes constructed for this triangle turn clockwise, while boxes for the triangle ACE turn counterclockwise.

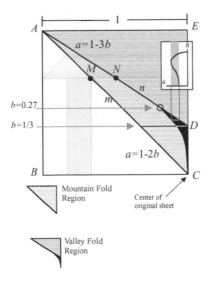

Figure 7. Diagram of the point P (quarter view).

6   Conclusion

The conclusion of my study is as follows: Even if we take point P anywhere in the sheet, we can construct the present type of twist boxes. The bottom square and the fold width can be determined only by determining point P.

The result of this study can be summarized in Figure 8. The square of the sheet is divided into several regions by the features of the boxes, that is, by the

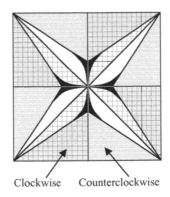

Clockwise   Counterclockwise

Figure 8. Diagram of point P (whole sheet).

clockwise or counterclockwise twist, by the mountain fold or valley fold, and by the single or double heights and angles. Figure 8 shows these regions. For the whole area of this sheet, the mountain fold twist boxes can be constructed. For the gray regions, a single valley fold box can be constructed for each point P. For the bill-like black regions, two boxes with different heights and twist angles can be made for each point P. And in the eight triangle regions divided by the diagonal or center lines, the clockwise and the counterclockwise twist boxes appear alternately.

The calculations of this study are based on those in [6].

References

[1]   Nagata, N., "Vessel", *Origami*, Nippon Origami Association, No. 217, (Sept. 1993) 10–11.

[2]   Nagata, N., "Dog box", *Origami*, Nippon Origami Association, No. 221, (Jan. 1994) 16–17.

[3]   Nagata, N., "Pyramid", *Origami*, Nippon Origami Association, No. 238, (Jun. 1995) 28–31.

[4]   Nagata, N., "Gift box with crane", *British Origami*, British Origami Society, No. 200, (Feb. 2000) 27–28.

[5]   Nagata, N., "Dokodemo box", in *Unit-Hiroba*, ed. by Fuse, T., Seibundou Shinkousha, (2001) 44–57.

[6]   Nagata, N., "A study on the twist in quadrangular origami tubes", *Proceedings of the Second International Meeting of Origami Science and Scientific Origami*, K. Miura ed., Otsu, Japan, (1994) 233–238.

To Fold or to Crumple?

B. A. DiDonna

1 Introduction

There is a reason the origami art-form attracts the attention of scientists and mathematicians. Origami is at once an expression of beauty, of order, and of the geometrical underpinnings of our world. The unique combination of a flat (two-dimensional) sheet in three-dimensional space yields an infinitely rich set of possible shapes and patterns, all derived from the basic elements of the fold and the conical point or vertex. The challenge is to find beauty within the simple inherent constraints that the paper sheet cannot be stretched or compressed in its own plane, but it can easily be bent out of its local plane.

When physicists are presented with a thin sheet of brightly colored paper, we are inclined to ask, "What would this sheet choose to do if I pushed on its edges?" Thus, we consider the ways that *nature* chooses to distort the sheet in response to external stimulous. A simple experiment is performed (see Figure 1). We crush the sheet between our hands, then pull it flat again. The result is clear: Nature also chooses to use folds and point vertices as preferred distortions of the sheet (though the results are much more chaotic than the efforts of an origami enthusiast). Why does this happen?

Once the question has been framed, we proceed to address it with our usual tools. We know that nature makes most of its decisions based on geometry and on energy. To describe the geometry of the sheet, we employ laws from differential geometry [1]. To describe the energy, we take Hooke's law for

(a)    (b)

Figure 1. A typical crumpled sheet. Image (a) shows a sheet of paper which has been lightly crumpled between the hands. Image (b) is the same sheet unfolded—lines and points resulting from plastic deformation show the former locations of folds and vertices in the crumpled state. Image courtesy the authors of [3].

linear elasticity [2]. We define the variables that are important to describe the sheet's response: The thickness of the sheet is labeled h, the length and width are labeled L (we assume they are the same order of magnitude), the energetic resistance to stretching is quantified by the Young's modulus Y, and the resistance to bending by the bending modulus κ. We move on to give plausible explanations for the whys and the what-ifs.

In this article, I will survey some of the results and insights physicists have gained into the crumpling phenomenon. In the next section, I show why nature chooses the fold as the preferred form of distortion. After that, I detail how this configuration lends strength to the crumpled sheet. Finally, I show how to make sense of the patterns in a highly crumpled sheet by viewing them as collections of folds. This article focuses on geometrical arguments while trying to avoid complex math. The interested reader can find more detailed information in referenced works.

2   Why Fold? Elasticity of a Flat Sheet

Let us consider a thin flat sheet of uniform elastic material. By "elastic", I mean that it can be stretched or compressed by some amount without breaking or damaging the material at all [2]. Most solids behave elastically over some range of distortions—even paper. For simplicity, we consider a perfectly elastic material, which can be distorted by any amount without being damaged. Paper is not perfectly elastic, since it can retain creases. However, the initial stages of crease formation are governed by elasticity, so the perfectly elastic approximation is appropriate at this point.

Elastic distortions in the plane of a thin sheet are parameterized by a quantity called *strain*. The strain on an object is the fractional change in its length away from its resting length. So, if a sheet of length L is stretched to length $L + \delta L$, the strain on the sheet is $\delta L / L$. We shall denote the strain by the symbol γ (see Figure 2(a)). Technically, the strain field is described by a 2×2 tensor, which gives the value of strain and shear in both directions on the sheet, but for our purposes there will be one clearly dominant component of this tensor, which we shall generically refer to as the strain. The strain energy of distortion of a sheet of thickness h is

$$E_S = YhA\gamma^2,$$

where A is the area of the sheet and the constant Y is the Young's modulus. The value of Y depends on the material.

The other way we can deform a thin sheet is by bending it. The energy required to bend a sheet into a cylinder of radius R is

$$E_B = \kappa A \left(\frac{1}{R}\right)^2,$$

where κ is the bending modulus and the quantity $C = 1/R$ is the *curvature* (see Figure 2(b)). Obviously, a thicker sheet is harder to bend and will thus

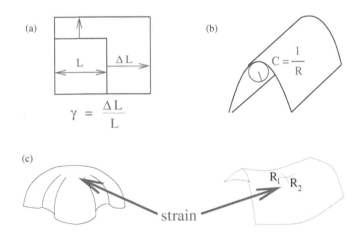

Figure 2. Strain and curvature. Image (a) illustrates strain in the plane of a sheet; (b) shows curvature of a sheet out of its material plane. The images in (c) demonstrate how simultaneous curvature in both directions on the sheet requires strain distortions in the sheet's plane.

have higher κ. It turns out that the bending modulus is related to the stretching modulus by

$$\frac{\kappa}{hY} \approx h^2.$$

This relation expresses the very simple fact that as a sheet of a given material, and thus a given Young's modulus Y, becomes very thin it is much easier to bend than to stretch.

With these facts established, there is only one other ingredient to consider: geometry. The strain and the curvature of a sheet are coupled by geometry. If an initially flat sheet is curved in one direction, then it must either remain flat in the other direction or it must be strained. Thus we cannot bend a flat sheet into a sphere or a saddle-shape without stretching it. This coupling is illustrated in Figure 2(c).

When we crumple a piece of paper between our hands, we are forcing it into a state where no direction is allowed to remain straight all the way from one end of the sheet to the other. Thus there must be places on the sheet which are bent in two different directions at once. The sheet therefore must contain some strain. But strain is very energetically costly, so the sheet will seek some way to fit into the shrinking volume between your hands with as little strain as possible. The best solution is for the sheet to confine most of its curvature to sharp folds, thus confining most of its strain to the points where these folds meet. The result is the typical collection of folds we see in a crumpeld sheet.

3 Closer to the Fold—Explaining Everything Else

The choice of folds is not the end of the story, however. We still would like to know how hard we must push to ball up the paper, or why the sound of crumpling is a collection of discrete ticks instead of a continuous roar. As it turns out, to better understand the behavior of a crumpled sheet, we must understand the shape of the sheet very close to the folds and vertices [3, 4, 5, 6, 7, 8].

The very arguments which favor the folding of the sheet also hint that there must be some interesting, nontrivial behavior very near the fold. After all, the sheet chose to fold in order to avoid paying energy for strain, but a sharp fold is locally like a cylinder of radius zero. So, the local bending energy cost for a sharp fold would be infinite. In real sheets, the region right around a fold will find a better energy balance by trading some strain to reduce its curvature. This balancing process is illustrated in Figure 3. Generically, when bending and stretching are traded against one another, the optimal balance will be found

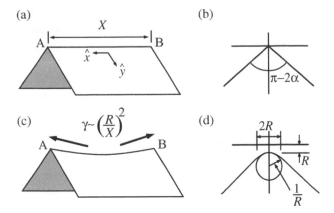

Figure 3. Local shape near a fold. Images (a) and (b) show the geometry of a sheet with one sharp fold. (a) is a side view, while (b) is a cross section of the fold at the midpoint of the sheet. Images (c) and (d) show how strain (lengthening) along the fold can be traded to soften the curvature across the fold. The geometry shown here, used for the arguments presented in [3], requires extensional strain along the midline—other geometries can have a net compressive strain along the midline, but the driving balance between curvature and strain remains the same [4].

when both energies are on the same order of magnitude,

$$YhA\gamma^2 \approx \kappa A \left(\frac{1}{R}\right)^2 \quad \Rightarrow \quad \gamma \approx \frac{h}{R},$$

where we have used the relation $\sqrt{\kappa/Yh} \approx h$. The details of the bend-strain balance in this case lead to an optimum energy when $\gamma \approx (h/X)^{2/3}$ and $C \approx h^{-1/3}X^{-2/3}$, where X is the length of the fold [3, 4]. This yields a total energy of about $\kappa (h/X)^{1/3}$, where we estimate the area of the fold to be about XR.

Putting it all together, most of the deformational energy in a crumpled sheet comes from counting up the $\kappa (h/X)^{1/3}$ contributions of all the various folds. The folds have much more energy than the relatively flat regions that surround them. Also, though they have much less energy *per unit area* than the high strain vertex points where folds meet, they also cover a much greater area of the sheet, and so outweigh these points in the end. So, we can get an accurate measure of how much energy is contained in a crumpled sheet by counting up the folds and assigning the appropriate energy to each based on its length using the formula above. This energy, plus the energy released as sound during the crumpling process, is the total energy required to crumple the sheet.

The sound of crumpling results from the nature of the buckling transition when one fold breaks into several. This transition has recently been investigated in detail by the author [9, 10]. After much work and very detailed simulations like those illustrated in Figure 4, it was shown that ridges buckle the same way as tin cans. This is real progress, since the buckling of tin cans and other thin cylinders has been studied in depth for the purposes of structural engineering [11, 12]. A cylinder, when compressed from end to end, can buckle in one of two ways. If it is very long and not too thin, it will buckle like a stick, with the whole structure blowing out to one side. Otherwise, it will buckle into the rippled pattern shown in Figure 5. The latter, rippled case also happens to be the way folds buckle when pushed at either end.

Having worked on this problem for so long, I can't resist adding a paragraph which is a little more technical:

The nature of the rippled buckling transition is a branch bifurcation— once enough force is applied to the cylinder, it is energetically favorable for this deformation to appear and grow. The reason that cylinders and tin cans buckle sharply, with the accompanying sound, is that the growth of this mode is very nonlinear. Although the flat, perfect cylinder shape is stable up to a

(a)   (b)

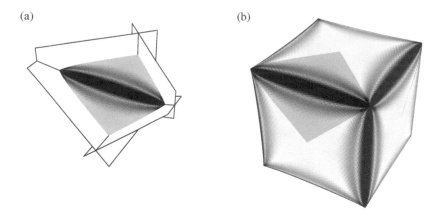

Figure 4. Typical elastic sheet used in numerical study. The resting configuration of the simulated sheet with no external forces. It also shows the reflection planes to which the sheet edges are constrained. (b) How the simulated sheet is equivalent to one edge of a cube, when the mirror images of the sheet across the reflective planes are drawn in. The thickness of the sheet is .0004 of the the edge length. Darker shading shows higher concentrations of strain energy. The entire simulated sheet is uniformly darkened to distinguish it from its mirror images in (b). The numerical grid is visible as a quilt-like texture.

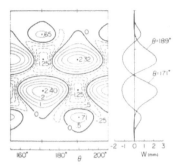

Figure 5. Diamond shaped buckling mode for a thin cylinder. This image was taken from [11]. On the right is a photograph of an aluminum cylinder buckled by application of straight downward forces at its ends. On the left is a relief map of a simulated sheet with the same buckling pattern.

high applied force, once this rippled pattern appears and starts to grow, the surface finds it can lower its energy even more by making the pattern grow larger. Thus, once the smooth surface is broken, the growth of the buckled shape is like an avalanche that doesn't stop until it has hit the bottom of its energetic mountain. Those readers familiar with bifurcation theory may have guessed that the transition was *subcritical* (i.e., the system jumped over an energetic mountain)—the surprising fact is that this is not the case (to continue the metaphor, this system climbs all the way to the top of the mountain, then jumps down).

The application of cylinder buckling theory to fold buckling is straight-forward. The folds are rounded across their middle, so we use the curvature of this bend in place of the curvature of a cylinder. Once again, we imagine force applied at either end of the fold. We put in all the dependencies for the point at which cylinders become unstable, and find that the ridges buckle when $\gamma \approx \frac{h}{R}$. This should be familiar from above, since we said that this condition already holds for ridges with no force applied. Thus we have shown that the energy scale for buckling is about the same as the energy scale for making the ridge in the first place. In fact, in a number of simulations, the author found that many different ridges buckled at an energy that was about 120% of their resting energy, regardless of the resting value of their energy. The energy loss at buckling was also on the same order.

4   What's Next

At this point, we have enough information to understand the shape of a crumpled piece of paper. We know its energy, and from the knowledge of how

one fold breaks into several, we can also estimate how much energy was released as sound while it was being crumpled. Though I have not discussed it,
we even have a bit of an idea why a new fold will appear in one place and
not another. Unfortunately, at this point in time, physicists have not worked
out the "big picture" of the crumpling phenomenon. Some of the important,
unanswered questions are how to predict the distribution of fold lengths in a
crumpled sheet, whether this distribution is universal for all crumpled sheets,
and how does the collection of folds in a crumpled sheet act together when
considered as a single object? Still, if we don't understand the forest, at least
we understand the trees.

Interested readers may be curious about generalizations of simple, flat sheet
elasticity. Progress has also been made in related fields like the blistering of
thin film coatings [13], the buckling of thin viscous sheets of fluid [14], and
the energetics of sheet-like molecules [15, 16].

References

[1] R. S. Millman, G. D. Parker, *Elements of differential geometry* (Prentice-Hall, Englewood Cliffs, N.J. , 1977).

[2] L. D. Landau and E. M. Lifshitz, *Theory of Elasticity* (Pergamon Press, New York, 1959).

[3] A. Lobkovsky, S. Gentes, H. Li, D. Morse, T. A. Witten, "Scaling Properties of Stretching Ridges in a Crumpled Elastic Sheet" *Science* 270, 1482 (1995).

[4] A. E. Lobkovsky, "Boundary Layer Analysis of the Ridge Singularity in a Thin Plate" *Phys Rev E* 53, 3750 (1996).

[5] M. BenAmar, Y. Pomeau, "Crumpled Paper" *P Roy Soc Lond A Mat* 453, 729 (1997).

[6] E. Cerda, S. Chaïeb, F. Melo, L. Mahadevan, "Conical Dislocations in Crumpling" *Nature* 401, 46 (1999).

[7] S. Chaïeb, F. Melo and J. C. Geminard, "Experimental study of developable cones" *Phys. Rev. Lett.* 80, 2354 (1998).

[8] E. Cerda and L. Mahadevan, "Conical surfaces and crescent singularities in crum-pled sheets" *Phys. Rev. Lett.* 80, 2358 (1998).

[9] B. A. DiDonna and T. A. Witten, "Anomalous strength of membranes with elastic ridges" *to appear in Phys Rev Let*, http://xxx.lanl.gov/abs/cond-mat/0104119.

[10] B. A. DiDonna, "Scaling of the buckling transition of ridges in thin sheets", *submitted to Phys Rev E*, http://xxx.lanl.gov/abs/cond-mat/0108312.

[11] N. Yamaki, *Elastic stability of circular cylindrical shells* (North Holland Press, New York, 1984).

[12] S. P. Timoshenko and J. M. Gere, *Theory of Elastic Stability* (McGraw-Hill Book Company, New York, 1961).

[13] M. Ortiz and G. Gioia, "The morphology and folding patterns of buckling-driven thin-film blisters" *J. Mech. Phys. Solids* 42, 531 (1994).

[14] A. Boudaoud and S. Chaïeb, "Singular viscous tablecloth" preprint.

[15] T. A. Witten, H. Li, "Asymptotic Shape of a Fullerene Ball" *Europhys Lett* 23, 51 (1993).

[16] L. A. Cuccia, R. B. Lennox, "Molecular Modeling of Fullerenes with Modular Origami" *Abstr Pap Am Chem S* 208, 50 (1994).

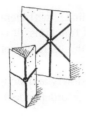

Folded Tubes as Compared to *Kikko* ("Tortoise-Shell") Bamboo

Biruta Kresling

1   Methods of Form-Finding

The present paper strives to define some rules for designing mechanically efficient lightweight structures for engineering purposes. It describes several patterns of folds in structures created from developable surfaces (i.e., surfaces formed by folding a flat sheet of paper with no embossing or significant plastic deformation). These structures are a special case of shell-structures [2]. In such structures, every point has Gaussian curvature equal to zero, which implies that there is a straight line tangent to the surface at every point of the surface.

Such a structure resists compressive loading when the load is applied along the surface in the direction of the straight lines. However, the surface is easily deformed from convex to concave (and vice-versa) when loaded perpendicularly to the straight line of the surface. Designing an efficient shell structure requires guiding the force flow along (and not perpendicular to) the straight lines. The efficiency of the design depends directly on the minimum of effort used to achieve this control of the force-flow. The mechanical behavior of the surface is thus directly dependent on the geometry of the surface.

To aid in finding suitable forms, the designer may be helped by studying mechanical principles in the morphology of natural structures. In the present paper, we examine tubular structures similar to features found in bamboo culms.

Natural structures commonly achieve states of minimum energy for two primary reasons:

Firstly, the evolutionary process in Nature tends to eliminate inefficient structures or structures that are poorly adapted to perform vital functions. It is possible to simulate this process by mathematical models like the "Strategy of Evolution" developed at the Technical University of Berlin by I. Rechenberg.

Secondly, in biological structures that build up their shapes slowly and step-wise, a feedback process occurs that makes growth and form interdependent. This relationship was pointed out in "On Growth and Form," a compendium by Sir d'Arcy Thompson [1] that became a classic reference. Thompson showed that the shape of living forms can be described by relatively simple mathematics. In many cases, simplified natural shapes and growth patterns can be modeled using the geometry of developable surfaces.

In our lab, research and experimental form-finding with developable surfaces started in workshops with students in industrial design and bionics in France, the UK and Austria. By means of crushing, twisting, and controlled local failure (using inserted mandrels), we induced spontaneous folding patterns at the point where the structure buckles or yields. Such patterns show a great regularity and the experiments can be repeated with similar results, which suggested to us that failure patterns in structures created from developable surfaces (i.e., sheets, corrugated surfaces, cylinders, cones, prisms, etc.) are minimum-energy configurations. We analyzed the patterns for their geometry and their related mechanics. Subsequently, the findings may be used in a conscious design process.

2   Geometry-Related Mechanical Properties of Normal Bamboo and of *Kikko* Bamboo Culms

Paper or cardboard models may mechanically simulate the principles within the growth patterns of bamboo culms. Such simulations help one understand the geometric and mechanical properties of the biological structure and—through comparison with engineering principles—help define optimization strategies for technical structures. This method is named "bionics" or "biomimetics" ([3], [4], [5]).

We compared normally grown bamboo culms with abnormally grown bamboo the so-called *kikko* (tortoise-shell) bamboo (Figures 1(a–d)), and modeled their mechanical behavior with paper structures. Normally grown bamboo culms are hollow and, like the culms of most grasses, consist of many nearly cylindrical segments (internodia) that elongate during growth between perpendicular diaphragms (nodia) (Figure 2(a)). The mechanical behavior of hollow culms and their stiffening diaphragms may be simulated by a paper structure with belt-like parts folded to form a diaphragm.

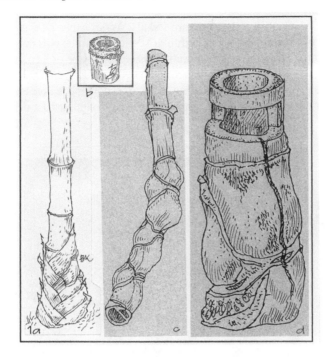

Figure 1. (a) Normally grown straight bamboo. (*Phyllostachys aurea*, China.) Base of a bamboo culm, with typical growth pattern of elongated internodia (walls) and perpendicular nodes. The length of the internodia increases with the height. Branches and leaves develop at the distal end of the culm. In (b–d) we see the sacred types of bamboo in Japanese tradition, used for the tea ceremony or in sake production. (b) Straight bamboo. One node of a straight bamboo, used as a lid rest for the tea ceremony, signed Furuta Oribe (1544-1615). (c) Abnormally grown so-called *Kikko* ("Tortoise-shell") bamboo. (Probably *Phyllostachys heterocycla f. pubescens*.) Bamboo used as a mixing stick for sake pulp. (d) Base of a bamboo in *kikko*-pattern, sculpted as a flower vase for the tea ceremony, by the master Sen-no-Rikyu (1522-1591) (b, d, Nomura Art Museum, Kyoto). (Drawings, B. Kresling).

3   Instructions for Provoking a Spontaneously Appearing "Kresling Pattern"

Fold lines build up coupled mechanisms. Whereas the classical origami process adds folds stepwise and approaches the final shape slowly, engineered folding design is based upon the comprehensive activation of all folds in a structure at a time. The "Kresling Pattern" may be obtained as shown in Figures 2 (b) and (c). This model creates a tube with folded diaphragms, perpendicular to the tube's axis, which stiffen the structure locally. Wrap paper or light cardboard tightly onto two mandrels of equal circular section. Pull the two mandrels apart in order to form a gap and twist coaxially. (The gap may be of any size, but take care that the height of the gap between the mandrels is less than the diameter of the wrapped paper tube). This twisting induces a spontaneous crease pattern on the segment corresponding to the gap; the pattern is similar to a diaphragm of

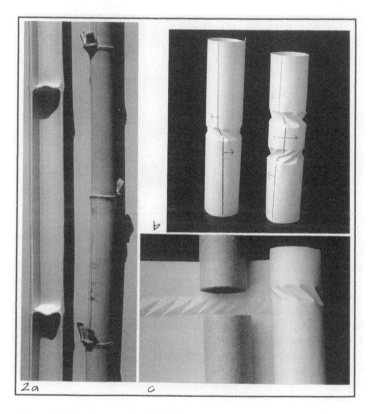

Figure 2. (a) Bamboo: cylindrical segments and perpendicular nodes. Normally grown bamboo culms are hollow and, similar to most other grasses, consist of many cylindrical segments (internodia) that elongate during growth and diaphragms (nodia) perpendicular to the segments. (b, c) Paper tubes with folded diaphragms as models of this growth pattern, "Kresling Pattern". A simple twisting of a paper wrapped on two mandrels, pulled apart in order to create a gap, creates in the region corresponding to this gap a spontaneous crease pattern similar to the diaphragm of a camera. The diaphragm locally stiffens the cylinder. The gap must not exceed a critical value, close to the diameter of the wrapped paper tube.

a camera. This pattern is defined in engineering as "shear-buckling". The tube still is able to carry axial loads, but the great difference is found in the vicinity of the diaphragm: The tube is now locally resistant to peripheral compressive load (applied in the direction perpendicular to the tube's axis). The experiment could be repeated in order to obtain multiple segments. It is easily observed that with increasing distance between the stiffening diaphragms, the tube decreases in resistance to ovalization.

This experiment was developed in courses on bionics with French design students and it was discovered that this pattern can be induced easily and quickly. We dubbed this and similar experiments "one-second-workshops". The pattern thus derived is known in engineering literature as a "shear failure" pattern, but when considering the mechanical properties of the diaphragm and of the applications of this simple but efficient mechanism (for example, for micro-tools for noninvasive surgery or space antennae), the pattern offers greater potential than is suggested by "a failure behavior under shear load".

For its proposals toward novel applications the pattern has been named "Kresling Pattern" (Poeppe in the German edition of Sc. Am., Sept. 2000). The paper model with stiffening diaphragms corresponding to the nodia of the straight bamboo culm shows that the tubular section that corresponds to the internodia is, however, still sensitive to bending and easily "ovalizes" (Figures 4 (a) and (b)). This, too, is a natural phenomenon. In natural bamboo, ovalization (due to excessive bending) induces shear strain and induces splitting. Splitting is the main cause of failure in bamboo.

In the case of abnormal growth, which occurs in so-called *kikko* ("tortoise-shell" pattern) bamboo, the mechanical problem of the internodia lacking bending stiffness is solved by a particular growth form (Figures 3(c) and 4(c)). In this bamboo, the internodia are short and bulged and the nodia are oriented obliquely to the stem with alternating inclination.

Kikko bamboo of the species *Phyllostachys heterocycla f. pubescens* grows in Japan in only a few places, for instance, in the area of Nishi Kyoku in Kyoto. *Kikko* bamboo is highly appreciated for its strength and canes are traditionally used as mixing sticks in sake breweries or as fishing rods. The great Japanese master of tea ceremony, Sen-no Rikyu, used *kikko* bamboo for a flower vase, for its expression of vigor and rusticity (Figures 1 (c, d)).

When carefully split lengthwise (parallel to the fibers, see Figure 3(a), obliquely grown nodes, thickened walls, and the wavy fiber direction of the abnormally short internodia reveal the secret of the *kikko* bamboo's strength:

> The force-flow passes without interruption from one outer wall through the oblique nodia to the wall of the other side. Three interwoven sinusoidal lines compose its shape.

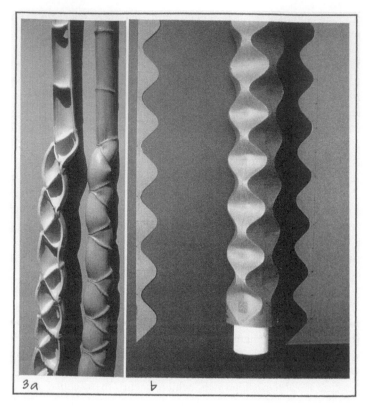

Figure 3. (a) *Kikko* Bamboo: bulged segments and oblique nodes. In the abnormally grown culms of *kikko* bamboo, the internodia are short, bulged, and the nodia are oriented obliquely to the length of the culm. The longitudinal section shows schematically three sinusoidal lines running from the wall of one side through nodes to the wall of the other side. (b) Paper structure with sinusoidal foldlines simular to the *kikko* pattern. Prototype of light tubular structures designed for carrying compressive, bending and torsional loads. The outer wall of the tube is reinforced by three pairs of crossed helical fold lines. Mechanically speaking, these tubes may be considered "high-mode buckled". (Design, B. Kresling and students at UTC, Compiègne, France).

When sectioned in the perpendicular plane, the walls are straight and the diaphragms are perpendicular to the axis of the culm—it presents again the ladder-like configuration similar to the normal, straight bamboo. Although *kikko* bamboo is appreciated for its structural strength and stiffness for some crafts, this bamboo is unable to reach full size, and is less competitive in nature in an environment of the taller, normally growing bamboo culms.

4 Shape Optimization for Folded Structures Derived from *Kikko* Bamboo

The *kikko* pattern provides the following principles for the design of high-mode buckled technical tubular structures, which are currently under research. (Figure 3(b) and Figure 5). Three important principles regarding design optimization of tubular structures are suggested by its shape.

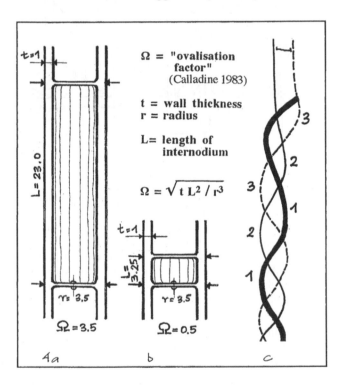

Figure 4. (a, b) Ovalization factor applied to cylinders with closed nodes. The length of a cylinder with closed caps is varied and calculated for its resistance to ovalization (after Calladine [2]). The cylinder corresponds to one internodium of a straight bamboo culm, the caps to the nodes. The relation of wall thickness/radius/length of the cylinder is given by the formula quoted above, where Ω is the dimensionless "ovalization factor". For a value 0.5 (or less) of Ω, Calladine indicates resistance to ovalization, for a cylinder with $\Omega=3.5$, exists a high risk of ovalization. The calculation does not take into account the anisotropy of the fibrous material. (c) Geometrical configuration of *kikko* bamboo with obliquely oriented nodes. The mean value of the segments is small, which indicates a relatively high resistance.

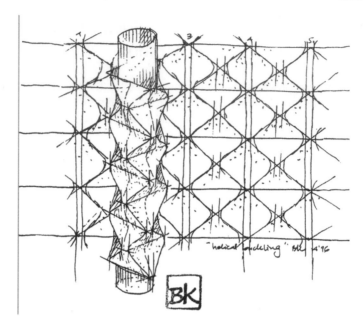

Figure 5. Folding pattern for the tubular structure. Sinusoidal fold lines are similar to the *kikko* pattern. The orthogonal grid corresponds to the definition of units (concave dimples). The oblique straight lines correspond to the crossed helical fold lines that wrap the structure. (One strip is overlapping and serves for glueing the figure into a tube.) To avoid problems of instability at the intersections of the helical fold lines, their intersections are smoothed into convex bridges and all fold lines are curved. The oblique grid serves as tangent lines for the tracing of the sinuosoidal fold lines. Note that the finished figure is elastic and that some loads might induce important shear stress.

Firstly, *kikko* bamboo shows in a longitudinal section an unbroken transition from one wall to the opposite wall, due to the oblique orientation of the diaphragms (Figures 3(a), 4(c)). This geometry increases the overall bending stiffness of the structure as compared to straight segments with perpendicular diaphragms. In the design of a tubular structure, this effect might be obtained either by walls that pass through the structure (and would compartmentalize it), or might be achieved by a series of stiffening activated fold lines that run helically on the outer surface and are crossed, so that the force-flow is virtually axial (which would not compartmentalize it). Figure 3(b) and its folding pattern (Figure 5) give one possible implementation, but there are many possible variants to it.

Secondly, *kikko* bamboo shows a mean value of relatively short segments. It is evident that the shorter the segments are, the better they resist ovalization. In computer simulations of the tubular structure that is considered as being the result of high-mode buckling, the frequencies (of the spatial oscillations) might be varied. It is premature to predict an "ideal" spatial frequency for the technical tube, but it is possible, for a first approach, to calculate the pattern corresponding to the dimensions of the biological shape (which is extremely stiff) and to virtually test the mechanical behavior after some geometrical variations.

Thirdly, *kikko* bamboo presents a pronounced asymmetry. One longitudinal section shows three interwoven sinusoidal lines, but cut in another way, the longitudinal section of the same structure shows a ladder-like configuration with straight walls and with diaphragms oriented perpendicularly to the walls. The mechanical response of such an asymmetric configuration to a compressive load is that of an elastic structure combined with an inelastic one. For the design of a tubular structure, a slight asymmetry (for instance, a slight twisting that changes the pitch of the helical fold lines) might help overcome the irregular buckling of the tube in response to compressive and/or bending loads.

At the moment, research by calculation and by computer simulation is under way at the engineering department and Centre for Biomimetics at the University of Reading, UK (in collaboration with A. Atkins and G. Jeronimidis).

In the final figures below, we describe some design rules that are based upon the particular mechanical properties of developable surfaces. The rules apply for the forms shown in this paper (Figure 6–8).

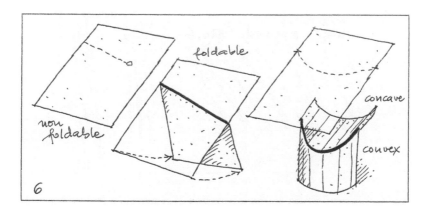

Figure 6. Geometry-related mechanical properties of folds in a developable surface. First mechanical property: A single fold line cannot terminate in a surface; it must run entirely across it and can only terminate on the border of the surface. Design consequence: A straight fold acts as a linear hinge (piano hinge). A curved fold automatically bends the adjacent surfaces, one side to concave, the other to convex.

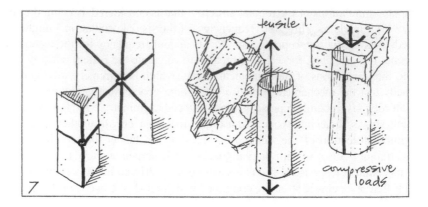

Figure 7. Geometry-related mechanical properties of folds in a developable surface. Second mechanical property: At any point of a surface, whether flat, bent, or crumpled, it is possible to draw a straight (i.e., uncurved) line. Design consequence: Force flow ideally travels on trajectories of straight lines. Those lines are loadable and can carry equally tensile or compressive forces.

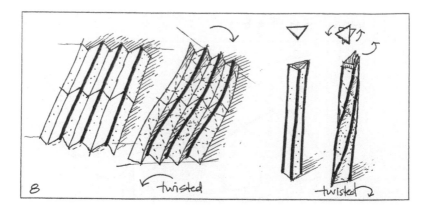

Figure 8. Geometry-related mechanical properties of folds in a developable surface. Third mechanical property: Both straight and curved fold lines are twistable. Twisting induces slightly bent neighboring surfaces and creates "relaxation points" (singularities) at regular intervals. Fold lines take the form of an "S" or a "?". Design consequence: By "overtwisting" or "overbending" fold lines, one may create a quasi-elastic behavior. Such shapes are self-regulating.

References

[1] Thompson, D'Arcy, *On Growth and Form*, New ed., Cambridge Univ. Press, (1963).

[2] Calladine, C., *Theory of Shell Structures*, Cambridge, Cambridge Univ. Press (1983).

[3] Kresling, B., "Plant 'design': Mechanical simulations of growth patterns and bionics", *Biomimetics* (J.F.V. Vincent and G. Jeronimidis, eds.), Plenum, New York 3, 3, (1996) 105–122.

[4] Kresling B., "Folded and unfolded nature", *Origami Science and Art. Proc. 2nd Int. Conf.*, (K. Miura ed.) Seian Univ., Otsu, Japan, (1997) 93–108.

[5] Kresling B., "Coupled mechanisms in biological deployable structures", *Proc. IUTAM Symp. Deployable structures: theory and applications* (S. Pellegrino ed.), Kluwer Ac. publ., Dordrecht, (2000) 229–238.

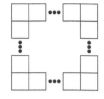

Square Cycles: An Introduction to the Analysis of Flexagons

Ethan Berkove and Jeffrey Dumont

1 Introduction

The creation of the flexagon will come as no surprise to anyone who has found himself or herself making random folds in a piece of paper for fun. Flexagons were created by an English graduate student named Arthur Stone in 1939 while studying at Princeton University. Since American and European paper sizes are different, Stone had to remove a strip from the end of his paper to get the sheet to fit into his binder. He folded the left-over strips into equilateral triangles, wove them into hexagonal shapes, and created the flexagon!

Flexagons are geometric objects with a mathematical flavor, so are an appropriate choice for a classroom exploration at a number of levels. In fact, the first author still remembers his first introduction to flexagons in a fourth grade math class (He recently found some of his models after an unusually thorough spring cleaning!). However, this accessibility does not imply that flexagons are easily understood objects. To the contrary, we have found flexagons to have an unusually rich structure. In this article, we hope to introduce the reader to some of the basic members of the flexagon family and give some indications of how one can go about and analyze them.

It's easy to fold a flexagon, and we strongly suggest creating your own as this experience is helpful in the rest of the paper (and it's fun). Start with a strip of nine equilateral triangles as in Figure 1 (called the *net* of the flexagon), and apply the following process:

Figure 1. Initial strip for trihexaflexagon.

1. Hold a in your hand.

2. Fold c over b.

3. Fold f over e.

4. Fold i over h.

5. Glue or tape a over the tab at the end of the strip.

Once completed, you should have the flexagon depicted in Figure 2. To begin flexing your new creation, mountain fold along the three radial edges that lie between two separate layers (the edges between triangles a and c, d and f, and g and i). At the same time, bring every other outside corner of the hexagon together to form the "imperial Y-Wing". Then open the flexagon at the middle.

Hint: If you color each face as it appears, you get a nice effect. You'll need three colors for this first case (not two!) which is why this flexagon is called a *trihexaflexagon*. There are many ways to represent each face: by color [10], by number [9], or by a silly face [6]. Find the face (a,i,g,f,d,c) read from left to right in a clockwise manner around the hexagon with a in the 11 o'clock triangle. If we flex up, we get the face (b,i,h,f,e,c). Performing another flex, we get the face (b,a,h,g,e,d), and a third flex gets us to (c,a,i,g,f,d). This is the same face we started with, although it has been rotated through an angle of $\frac{\pi}{3}$ radians. We can summarize all this information in what is known as a *structure diagram* as seen in Figure 3.

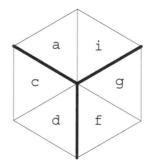

Figure 2. The trihexaflexagon.

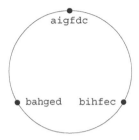

Figure 3. The structure diagram for the trihexaflexagon.

Permutations of the faces of a flexagons are known as *symmetries* of the flexagon. The argument above can be extended as in [6] to show that the symmetry group of the trihexaflexagon is D_{18}, the dihedral group with 36 elements.

Next, we would like to see if we can do a similar analysis of the hexa-hexaflexagon. The hexahexaflexagon can be constructed from the strip of triangles in Figure 4. To create the flexagon:

1. Wrap the strip. In other words, fold a over b, c over d, etc. The result is a single strip of pairs of triangles as in Figure 1.

2. Fold this strip as in the case of the trihexaflexagon.

3. Glue or tape the tab over a.

It's easy to find a flex that when applied to the hexahexaflexagon three times has the same effect as rotating the flexagon through an angle of $\frac{\pi}{3}$. However, in this case, every radial edge lies between two separate layers, so there are six possible initial flexes to choose from. This complicates the structure diagram considerably. By carefully tracing through the faces, one finds the following:

1. There is a main cycle, where flexes are possible at every radial edge.

2. There are three secondary cycles, where once into the cycle, the flex position is uniquely determined.

3. Any flex applied three times rotates the flexagon by $\frac{\pi}{3}$.

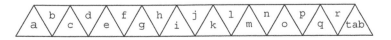

Figure 4. The initial strip for the hexahexaflexagon.

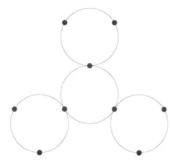

Figure 5. The structure diagram for the hexahexaflexagon.

This information can be easily summarized in the structure diagram in Figure 5. There appear to be nine vertices, i.e., nine faces in the hexahexaflexagon, but if one colors the faces as suggested by [10], three faces show up twice in the structure diagram, hence the flexagon's name. We note, however, that if you also take into account the number of sheets under each triangle, then in fact the flexagon has nine separate faces.

We also note that by increasing the number of triangles in the initial strip to 36, one can construct the nonahexaflexagon. The construction is similar to that of the hexahexaflexagon. Wrap the strip of triangles once to get the initial strip for the hexahexaflexagon, wrap again to get the initial strip for the trihexaflexagon, and do a final wrap to finish the flexagon. The resulting flexagon should have nine faces. We leave it to the reader to find the resulting structure diagram (which will have 21 vertices).

The hexaflexagon family is very well understood. [8] and [10] give a nice classification of all hexaflexagons by their structure diagrams. [9] introduces the concept of the *pat*, the thickness of one slice of the hexagon, which is used both to determine acceptable pat structures for flexagons and to enumerate the possible number of flexagons built from a given number of triangles. [2, 3, 7] provide lengthy references to other work on flexagons. More recently, Hilton, Pederson, and Walser studied the symmetries of trihexaflexagon in [6] and identified its symmetry group as D_{18}.

2   The Tetraflexagon Family

It is a reasonable question as to what other types of flexagons are possible. Probably the closest relatives to the hexaflexagons are the flexagons built from squares, the "tetraflexagons". We first learned of these objects in an article by Martin Gardner [4], where he discussed some of the easiest examples of tetraflexagons. It also appears that Stone and his colleagues knew about the tetraflexagons, but were unable to make any progress in classifying their structure [4]. With all that is known about the hexaflexagons, we initially felt that this family should be easy to analyze. We now feel that the tetraflexagon family is, if anything, richer in structure than its better known hexaflexagon relatives.

It is pretty clear that the simplest net, a $1 \times n$ strip of squares, doesn't make much of a flexagon at all, so any flexagon must have a net with corner squares. In fact, we will assume that our nets have precisely four corner squares and that the finished flexagons have 180 degree rotational symmetry about an axis perpendicular to the plane of the flexagon. In order to fold into a 2×2 square, nets can come in two different forms: $m \times n$ rings with both m and n even, and $m \times n$ "lightning bolts" with at least one of m or n odd (see Figure 6). This is easily confirmed by construction and again, we recommend the reader try some examples on his or her own.

The simplest way to fold tetraflexagons is to start at one end of the net and to roll the strip of squares repeatedly over or under until a corner is reached. Then turn the model 90 degrees and repeat the process, making sure to keep the flexagon symmetric. When the last square is reached, tape the final edge to the proper edge on the first square. We found it helpful to rotate the flexagon 180 degrees and peek at the rolled side in order to determine which edge to tape to which at the final step.

The resulting flexagon should look similar to Figure 7(a). In order to flex, find an partially free flap on the front side that can be lifted off the flexagon as in Figure 7(b). Fold the flexagon in half so the free flap is out. If you folded the tetraflexagon correctly, there will be a similar free flap kitty-corner to the one in front. Once the flexagon is folded in two, grab the two free flaps and open the flexagon; you should see a new face of the flexagon as in Figure 7(c).

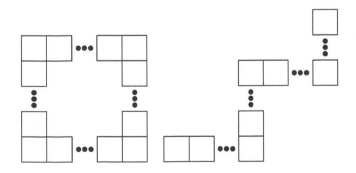

Figure 6. Rings and lightning bolts.

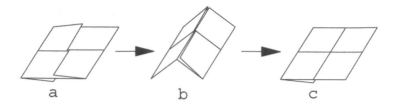

Figure 7. Flexing the tetraflexagon.

In some sense, both the front and back edges are simultaneously being rolled or unrolled. We distinguish four possible types of flexes: flexes up or down in the vertical directions ("book open" and "book closed"), and flexes up or down in the horizontal direction ("laptop open" and "laptop closed").

Tetraflexagons constructed in this way can be completely classified. Sample structure diagrams, for the case of 4 × 4 rings, are shown in Figure 8. These structure diagrams are L-shaped, with the length of each leg of the "L" determined by the number of squares in the straight strips that make up the net. Arrows in the diagrams correspond to like flexes. In the horizontal direction, arrows point in the direction of a book open flex (or book closed flex in the mirror image flexagon). In the vertical direction, arrows point in the direction of a laptop open flex (or laptop closed flex).

However, tetraflexagon structure diagrams can be considerably more complicated than the structure diagrams above suggest. Consider the 3 × 3 lightning bolt in Figure 9. Fold square b over square c, square d over square e, and square g over square f. Then tape square h to square a along the right edge. The resulting flexagon, which is also the simplest possible tetraflexagon (in terms of the number of component squares), has a four cycle for a structure diagram! This is the tetraflexagon analog of the trihexaflexagon.

It is possible to build tetraflexagons with a more complicated structure diagram by starting with nets larger than the 3 × 3 lightning bolt. For example, start with a 4 × 4 ring. Instead of rolling the straight strips of squares, make

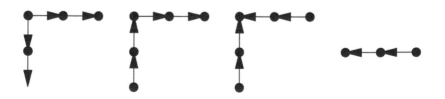

Figure 8. The structure diagrams of the "L"-shaped tetratetraflexagon.

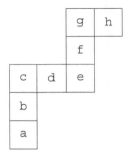

Figure 9. Initial strip for a tetraflexagon.

a single fold on one side of the first and third corner squares. This will yield a 4×3 lightning bolt. This process can be repeated until the 2×2 final configuration is reached. The folder, however, must take care the final flexagon is symmetric. Listed in Figure 10 are the structure diagrams for what we believe are all flexagons that can be made from a 4×4 ring. Although most of the diagrams are fairly basic, some folding patterns result in both cycles and straight sections in the structure diagram. And one folding pattern even results in two linked cycles. That such variation can occur in even this simple case suggests to us that the general structure diagrams for tetraflexagons can be quite complicated. To date, we have not been able to determine a solid relationship between the folding pattern of a tetraflexagon and its structure diagram.

3 Conclusions

Flexagons have been around for three quarters of a century, but these easily constructed mathematical recreations continue to be worth studying. The study of the tetraflexagon family, in particular, contains numerous open questions:

1. Is there an easily described connection between folding patterns and structure diagrams for the tetraflexagon family?

2. Are there certain configurations that are impossible in a structure diagram? For example, we believe that a cycle with four secondary cycles, analogous to the structure diagram for the hexahexaflexagon, is impossible.

3. How many possible different tetraflexagons can be folded from a given net?

4. What tetraflexagons are possible if one allows the nets to contain more than four corner squares?

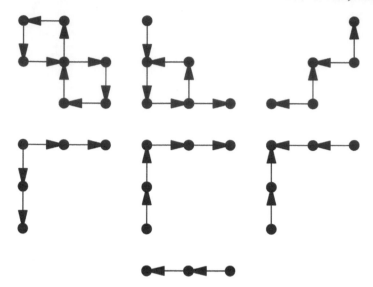

Figure 10. Structure diagrams from a 4×4 ring.

There is a good deal of material available from the analysis of the well-known hexaflexagon cases, much of which can be applied to the analysis of tetraflexagons. Furthermore, flexagons are accessible–a student can pick up the basics of flexagons within a day, and read most of the available literature with little additional background. We believe that flexagons are a mathematical object with great potential. They will be of interest to students of mathematics from the point of view of an unusual addition to a math classroom all the way to a open-ended research topic and a first exploration into mathematical research.

References

[1] P. B. Chapman, Square Flexagons, *The Mathematical Gazette*, Vol. 45, (1961), 192–194.

[2] M. Gardner, *Hexaflexagons and Other Mathematical Diversions*, University of Chicago Press, Chicago, 1988.

[3] M. Gardner, *The 2nd Scientific American Book of Mathematical Puzzles and Diversions*, Simon and Schuster, New York, 1961.

[4] M. Gardner, About Tetraflexagons and Tetraflexigation, *Scientific American*, May 1958, 122–126.

[5] M. Gilpin, Symmetries of the Trihexaflexagon, *Mathematics Magazine*, 49, No. 4 (1976), 189–192.

[6] P. Hilton, J. Pederson, and H. Walser, The Faces of the Trihexaflexagon, *Mathematics Magazine*, 70, No. 4 (1997), 243–251.

[7] J. Madachy, *Mathematics on Vacation*, Charles Scribner, New York, 1966.

[8] T. O'Reilly, Classifying and Counting Hexaflexagrams, *Journal of Recreational Mathematics*, 8, No. 3, (1975), 182–187.

[9] C. Oakley and R. J. Wisner, Flexagons, *American Mathematical Monthly*, 64 (1957), 143–154.

[10] R. F. Wheeler, The Flexagon Family, *The Mathematical Gazette*, 42, No. 339 (1958), 1–6.

Self-Assembling Global Shape Using Concepts from Origami

Radhika Nagpal

1   Introduction

It is fascinating how global phenomena appear from the local interactions of simple individuals with simple rules and no global intent or understanding. Colonies of ants and termites cooperate to achieve global tasks and build complex structures. Even more astounding is the precision and reliability with which cells with identical DNA cooperate to form complex structures, such as ourselves, starting from a nearly homogeneous egg. Developments in microfabrication and MEMs devices are making it possible to bulk manufacture tiny computing elements integrated with sensors and actuators. It will become feasible to create programmable reconfigurable materials and intelligent structures by embedding millions of these elements into materials. Understanding how global behavior and shape emerge from myriad simple parts will be critical to making intelligent materials a reality, as well as understanding developmental biology.

There have been a variety of approaches to understanding such global phenomena. Cellular automata and artificial life use an empirical trial-and-error approach to finding local rules, but provide no framework for engineering local rules for a specific goal [10]. Genetic algorithms and evolutionary approaches generate local rules without any understanding of the individual element behavior, making them difficult to verify and characterize. Approaches within the applications community have therefore focused on centralized control- and search-based approaches [2, 13]. My work is part of a larger vision called Amorphous Computing [1], whose goal is to identify engineering principles for designing robust self-organizing that achieve pre-established goals.

In this paper, I present the Origami Shape Language (OSL) for instructing a sheet of identically programmed, flexible, autonomous cells to assemble themselves into a predetermined global shape. The global shape language is inspired by Huzita's axioms of origami [4]. The cell programs and sheet model are inspired by studies of developmental biology and epithelial cell morphogenesis [9, 8]. The self-assembly works as follows: Using the Origami Shape Language, a desired global shape is expressed as a folding sequence on a continuous sheet. This description is compiled into a program for a single cell. A cell has limited resources and can only communicate with a small number of nearby cells. A sheet consists of a large number of identically programmed cells with a few initial conditions. The cells execute the program and coordinate to deform the sheet into the desired global shape.

With this language, one can reliably self-assemble a wide variety of complex predetermined shapes and patterns, using local interactions between cells. Several examples are presented: flat origami shapes, a Euclidean construction, a tessellation pattern, and a biological caricature. The expressiveness of the language is derived from the descriptive power of the origami axioms. The cell program is automatically compiled from the global shape description, which is in contrast to most cellular automata and artificial life research. The compilation process takes advantage of the fact that each origami axiom can be translated into a local rule, and the final shape can be expressed as a sequence of known local rules.

The language provides important insights into the relationship between local and global descriptions of behavior. For instance, origami is "scale-independent", which implies that the folding sequence describes a shape without reference to size. The size of the shape is determined by the size of the initial sheet. Scale-independence is also seen in biological systems, but there is very little understanding as to how it might be achieved. The origami shape language not only provides insight into how cells can achieve scale-independence, but also how the same cell program can create many related shapes without modification.

2 A Programmable Cell Sheet

Imagine a flexible substrate, consisting of millions of tiny interwoven programmable fibers, that can be programmed to assume a large variety of global shapes. One could design many complex static and dynamic structures from a single

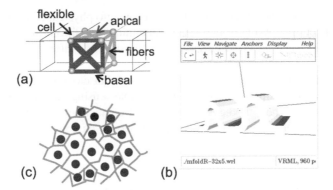

Figure 1. (a) Flexible cell model. (b) Three-dimensional dynamic simulation of a folding sheet. (c) Randomly distributed cells.

substrate. For example, a programmable assembly line that moves objects by producing ripples; manufacturing by programming flexible substrates; reconfigurable structures for deploying in space, that fold compactly for storage, but then unfold into some other structure.

Morphogenesis (creation of form) in developmental biology can provide insights for creating programmable materials that can change shape [11]. Epithelial cells, in particular, generate a wide variety of structures: skin, capillaries, and embryonic structures (gut, neural tube), through the coordinated effect of many cells changing their individual shape.

Our model for a flexible programmable material is inspired by epithelial cell sheets—the programmable material is composed of identically programmed, connected, flexible cells that create complex structures through the coordination of local shape changes in individual cells. The model for folding is adapted from a model by Odell *et al.* of epithelial cell folding during neurulation in embryos [9]. An epithelial cell can change its shape by actively contracting the fibers in its apical (top) and basal (bottom) membranes (Figure 1(a)). Many cells can coordinate to fold the sheet into a complex structure, such as Figure 1(b). Our model for the sheet is aimed at exploring shapes that can be created through folding along straight lines. Many cells in a line can coordinate to fold the sheet. The line must have many cells along the width and cells must be able to determine through local communication which fibers to contract. However, with this one simple mechanical operation, i.e., a flat fold, many different shapes can be achieved.

Computational Model for a Cell: All cells have the *identical* program, but an individual cell executes this program autonomously based on local communication. A cell can communicate only with a small local neighborhood of cells (within a distance r) or with cells that come into physical contact with it as a result of folding. Aside from a few simple initial conditions and apical/basal polarity, cells have no knowledge of global position or interconnect topology. Individual cells have limited resources and no unique identifiers; instead they have random number generators. There is no global clock or external beacon for triangulating position. Cells have very simple sensing and actuator control; a cell can sense when another cell is in direct contact with its apical or basal surface, and a cell can attempt to contract fibers in its apical or basal surface along a locally determined orientation.

3   The Origami Shape Language

Origami can be considered as a language for constructing global shape from a continuous sheet. The construction tools are a set of folding techniques. Dr. Humiaki Huzita has described a set of six axioms for constructing origami folds [4]:

1. Fold a crease between two points $p1$ and $p2$.

2. Given two points $p1$ and $p2$, fold $p1$ onto $p2$ (perpendicular bisector of the the line $p1p2$).

3. Given two lines $L1$ and $L2$, fold $L1$ onto $L2$ (bisector of the angle between $L1$ and $L2$).

4. Given $p1$ and $L1$, fold $L1$ onto itself through $p1$ (line perpendicular to $L1$ through $p1$).

5. Given $p1$ and $p2$ and line $L1$, make a fold that places $p1$ on $L1$ and passes through $p2$.

6. Given $p1$ and $p2$ and lines $L1$ and $L2$, make a fold that places $p1$ on $L1$ and $p2$ on $L2$.

The Origami Shape Language (OSL) describes global shapes that can be constructed using the origami axioms. Figure 2 shows how the construction of an origami cup can be expressed as an OSL program (Figure 3). The basic elements in the language are points and creases. Initially the sheet has only boundary conditions: points $c1$–$c4$ and edges $e12$–$e41$. The axioms generate new creases from existing points and creases. New points are generated by the intersection of previous creases. Because not all creases are meant to be folded, the fold execution is separated from the axioms. Also a fold may not go through

```
(define d1 (axiom2 c1 c3))
(execute-fold d1 basal landmark=c3)

(seepthru #t)
(define d2 (axiom3 e12 d1))
(define p1 (intersect d2 e41))
(define d3 (axiom2 c2 p1))
(execute-fold d3 apical landmark=c2)

(define p2 (intersect d3 e12))
(define d4 (axiom2 c4 p2))
(execute-fold d4 apical landmark=c4)

(seepthru #f)
(define l1 (axiom1 p1 p2))
(execute-fold l1 apical landmark=c1)
(define l2 (axiom1 (intersect d2 e34)
      (intersect d3 e23)))
(execute-fold l2 basal landmark=c3)
```

Figure 2. The OSL program for an origami cup.

all layers. If seepthru is true the fold goes through all layers, otherwise the fold acts as if the sheet were flat. Extensions to the language allow folds though multiple layers (not presented here). In origami diagrams, there is an implicit top surface. A valley or mountain fold is relative to this top surface and a new surface must be chosen after the fold. The top surface is made explicit in OSL by having the sheet maintain an apical (top) and basal (bottom) surface. There are two types of folds, apical and basal, that correspond to the valley and mountain folds. Given a crease, a fold that puts the apical surface on the inside is an apical fold and vice versa. After a fold is executed, the apical surface must be re-determined. The landmark in execute-fold specifies which side of the crease will move and hence which side of the sheet will reverse its apical/basal polarity. The choice of landmarks is important. In the case of the cup, the choice of landmarks ensures that both lateral flaps of the cup end up on the same side. All folds are flat simple folds, which implies that the structure always folds flat.

3.1 Why Origami?

The most attractive feature of origami is that one can construct a wide variety of complex shapes using a few axioms, simple fixed initial conditions, and one mechanical operation (a fold). Origami has considerable descriptive power.

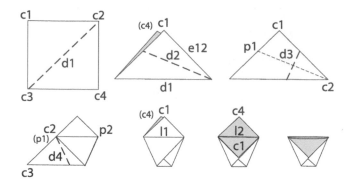

Figure 3. Folding sequence for an origami cup.

Huzita has proven that the axioms can construct all plane Euclidean constructions, and also solve polynomials of degree three, such as cube doubling and angle trisection [4]. Lang has shown that all tree-based origami shapes can be automatically generated by computer, revealing relationships to disk packing [7]. Demaine *et al.* provide a method for constructing scaled polygonal shapes [3]. Not all of these can be expressed using Huzita's axioms, however there is a large literature of shapes that can [5]. As the relationship between origami constructions and geometry is explored further, the results will directly impact this work.

This suggests an advantage to using a *constructive* global description. Rather than trying to map the desired goal directly to the behavior of individual elements, the problem is broken up into two pieces: (a) how to achieve the goal globally, and (b) how to map the construction steps to local rules. Thus we can take advantage of current understanding in other disciplines of how to decompose a problem.

4   The Cell Program

The OSL shape program is compiled into a cell program. All cells execute the same program and cells differ only in a small amount of local dynamic state. The compilation process takes advantage of the constructive origami description. Each global operation is translated into a local rule. Each local rule is composed from a small set of biologically inspired primitives.

Gradients: Gradients are analogous to chemical gradients secreted by biological cells—the concentration provides an estimate of distance from the source of the chemical. For instance, in the drosophila embryo, two different chemicals are emitted from opposite ends of the embryo and are used by cells to determine whether they lie in the head, thorax, or abdominal regions [8]. A cell creates a gradient by sending a message to its local neighborhood with the gradient name and a value of zero. The neighboring cells forward the message to their neighbors with the value incremented by one and so on, until the gradient has propagated over the entire sheet. Because cells communicate with only physically nearby cells, the gradient provides an estimate of distance from the source.

Epithelial Cell-Folding: This primitive is derived from the mechanical model of an epithelial cell. A cell can decide when to fold.

Cell-to-Cell Contact: This is how cells sense changes in the environment. A cell can detect when another cell comes into physical contact with its apical or basal surface as a result of a fold. Cells can communicate with other cells through direct contact. As a result, gradients can *seep through* cells in contact, allowing many layers of the sheet to act as a single thick layer.

Polarity Induction: All cells start out with the same initial apical/basal polarity, but this is not fixed. Cells can invert their internal polarity when induced to do so.

4.1 Composition into Local Rules

Points and creases in OSL correspond to boolean variables in the local state of a cell. An OSL point or crease is represented by a group of cells whose corresponding state variable is set to true. There are no leaders of the points or creases; no one cell is in charge of the group. Initially, all cells have boolean state variables for $e12, e23, e34, e41$ and $c1, c2, c3, c4$. If a cell is near the edge $e12$ of the sheet, then the corresponding state variable is true, and so on. The initial conditions automatically break the symmetry of the sheet, therefore we are not worried about symmetry breaking. However, the initial conditions are very simple; cells do not know where they are within an edge and the remainder of sheet is homogeneous, just like a sheet in origami.

This section describes how each of the global OSL operations can be implemented as a simple cell program (also called local rule) using the primitive behaviors. The global operation is achieved by all cells executing the same local rule based on their internal state.

The axioms use gradients to determine which cells belong to the crease. Axiom 1 uses tropism to create a crease from point $p1$ to $p2$. Tropism implies the ability to sense the direction of a gradient by comparing neighboring values; Coore introduces local programs based on tropism [1]. $p2$ creates a gradient and $p1$ grows a crease toward $p2$ by following decreasing gradient values (Figure 4(a)). Axiom 4 re-uses the same local rules to grow a crease from point $p1$ to the crease.

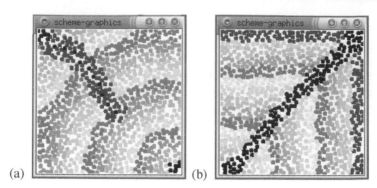

(a)   (b)

Figure 4. (a) Axiom 1 in progress, using tropism to generate a crease from corner $c1$ to $c3$. (b) Axiom 3 crease, the crease cells are equidistant from edges $e12$ and $e23$.

Axiom 2 creates a crease line such that any point on the crease is equidistant from points $p1$ and $p2$. Therefore, if $p1$ and $p2$ generate two different gradients, each cell can compare if the gradient levels are approximately equal to determine if it is in the crease. Axiom 3 uses the same local rules as Axiom 2. The difference is that the gradients are produced by creases rather than points (Figure 4(b)).

Cell-to-cell contact and gradient seep through allow the same local rules to apply within the context of the folded sheet. The local rules for an axiom can be expressed as a simple cell program. In Figure 5, the local rule for Axiom 2 is expressed as a Scheme procedure. Currently Axioms 1–4 are implemented, but Axiom 5 can also be implemented using tropism. Each

```
(define (axiom2-rule p1 p2 g1 g2)
  ; arguments are boolean state p1 p2
  ; and gradient names g1 g2

  (if p1 (create-gradient g1))
  (if p2 (create-gradient g2))
  (wait-for-gradients g1 g2)

  ; generates a crease of width apprx 2r
  (if (< (abs (- g1 g2)) 1)
    #t
    #f))
```

Figure 5. Local Rule for Axiom 2.

axiom attempts to produce creases that are approximately twice the width of the local communication distance r.

To execute a fold, the cells in the crease would actively contract their apical or basal fibers. The abstract sheet simulator approximates this by performing the fold along the best-fit line of the crease cells. In order to redetermine the apical surface, the landmark cells create a gradient before folding that cannot pass through the crease. The gradient marks all the cells on one side of the crease. After the fold, these cells reverse their polarity. Thus the sheet constantly maintains an apical and basal surface. This idea is inspired by tissue induction, where one region of cells induces its type or function on other cells through contact.

4.2 Compilation of the Cell Program

The OSL shape program is compiled into a sequence of calls to the appropriate local rules. Each local rule takes the current state of the cell and returns a true or false value depending on whether the cell belongs to the new point or crease. Each local rule creates new gradients, therefore the process is very communication intensive. However, the use of a gradient is short-lived, therefore a cell need not keep track of more than three gradients at a time. In general, the amount of local state per cell is very small: a boolean per distinct point or crease name and space for three gradient values. Furthermore, the majority of the cell program is conserved and re-used across all shapes. The majority of the code implements intercellular communication, the primitives, and the local rules; the sequence of calls corresponding to the shape forms a small part.

5   Simulation Examples

Figure 6 shows different shapes and patterns generated by compiling a global description in OSL and simulating the execution of the cell program on thousands of cells, where each cell communicates with only 15 to 20 neighboring cells on average. Figures 6 (a–f) show some of the stages during the formation of an origami cup on 4000 cells. The cell program was generated by compiling the code from Figure 2. Figure 6(g) shows the final crease pattern of the cup after unfolding the sheet. In general, one can create any flat origami that is composed of simple folds. Figure 6(h) shows a partially folded origami airplane using 6000 cells. Figure 6(i) shows a rouge container as described in [5]. One can also program two-dimensional patterns using OSL. Huzita's axioms can generate all straight edge and compass diagrams, and thus theoretically, we can self-assemble all such patterns. This is an extremely powerful idea because it can be used to systematically self-organize regular patterns such as grids or triangulations as well as nonregular patterns such as the CMOS inverter in Figure 6(j). Folding can also be used to create symmetric patterns by folding the sheet into a tile and creating lines that seep through, and then unfolding the pattern (Figures 6 (l, k)). This is a similar idea to creating snowflakes by cutting folded paper.

6 Scale-Independence and Related Shapes

The formation of the same structure at many different scales is common throughout biology. Internally complex organs, such as lungs and kidneys, exist in species of widely different sizes. A fragment of hydra one hundredth the volume can give rise to an almost complete animal, and sea urchin embryo develop normally over an eightfold size difference. Many of these cases would suggest that the processes for forming morphology are capable of scaling to different sizes, without modification of the DNA. However, biologists currently have very little understanding of how this is achieved.

The Origami Shape Language is in fact scale-independent. The same global program (and hence the same cell program) can produce a shape at different scales. Figure 7 shows cups formed from 2000, 4000, and 8000 cells, all using the *same* program as in Figure 2. Only the crease width is determined by the local communication radius, the global shape scales with the number of cells.

At the cell level, scale independence is achieved by using relative comparisons of gradients. At the global level, scale-independence is a side-effect of using origami. The sheet starts with only a boundary. Each operation (fold) is relative to the current boundary and creates new boundaries. Thus complexity is generated by recursively applying simple operations, while still remaining relative to the original boundary. Thus we are able to create highly complex structures without reference to size.

This leads to an interesting observation, which is that by changing the *shape* of the boundary, we can also change the shape generated. Hence if we start with a kite-shaped paper, we can generate fat short cups or tall thin cups with no modification of the program. The idea of shapes in biology being related by coordinate transforms was one of the many fascinating theories presented by D'Arcy Thompson in his book *On Growth and Form* [12]. He observed how the forms of many related animals (i.e., crabs, fish, skulls) could be plotted on Cartesian coordinates and transformed into one another by stretching along different axes. He claimed that these relationships extended not only to superficial characteristics, but also internal organs. Although he did not investigate the mechanisms for the transformations, he suggested different growth rates of cells along different axes as a mechanism that could produce such patterns. The

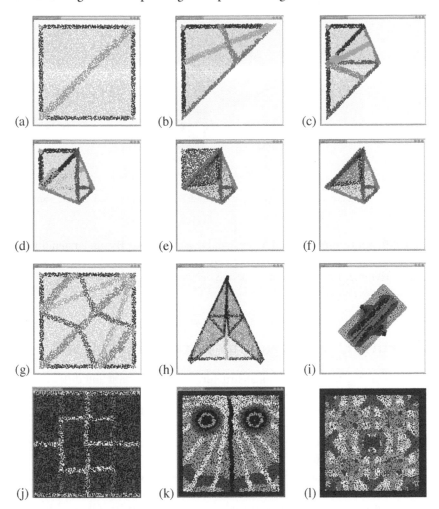

Figure 6. (a–f) Sequence showing 4000 cells folding into a cup using the cell program compiled from the OSL program in Figure 2. (g) Final crease pattern after opening the cup. (h) Partially folded origami airplane. (i) Rouge container. (j) CMOS inverter pattern. (k) Moth wing pattern caricature. (l) Tessellation pattern.

Origami Shape Language suggests another mechanism—changing the shape of the initial boundary.

Independently, origamists have been studying related shapes. At this same conference Dr. Kawasaki showed that the *orizuru*, the traditional origami crane, can be created from rhombus and kite-shaped paper, resulting in related *orizuru*

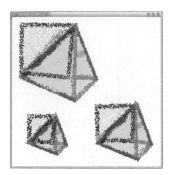

Figure 7. Origami cups with 2000, 4000, and 8000 cells, created by the same cell program.

with differently proportioned wings, tails, and heads. By understanding the essence of the folding sequence one can even create *orizuru* from rectangular paper (asymmetrical wings) or recursive *orizuru* (*orizuru* embedded in the wing of a larger *orizuru*) [6].

7   Future: Extension to Three-Dimensional

Self-assembling three-dimensional structures is an important and difficult problem. Given the current sheet model, one cannot fold precise angles because the individual cells have no means of measuring (no feedback) what angle has been attained. Flat folds do not require precision and are therefore reasonable. Either some sort of feedback must be added to the cell or the structure must not require precise angles.

Here, again, origami can play a role. Paper is also incapable of holding precise angles. Therefore, three-dimensional structures in origami, such as boxes and polyhedra, are constructed as *self-stabilizing structures*. Many of these structures use a combination of flat folds and precreases. These structures rely on the inherent stiffness of paper, which means that unbroken facets of the sheet act like rigid plates. By precreasing, the stiffness is broken, effectively creating a *hinge*. Thus the three-dimensional structure is composed of flat folds and hinges. An example is the corner module that creates a right angle corner—an origami box uses this crease pattern at each corner. Origami may also take advantage of tucking whereas the cell sheet may take advantage of adhesion between cells. Currently, three-dimensional origami exists as a collection of example cases—achieving a more general theory for how to construct self-stabilizing structures could play a very important role.

References

[1] H. Abelson and D. Allen and D. Coore and C. Hanson and G. Homsy and T. Knight and R. Nagpal and E. Rauch and G. Sussman and R. Weiss, "Amorphous computing", *Communications of the ACM*, Vol. 43, no. 5, (2000).

[2] A. Berlin, *Towards Intelligent Structures: Active Control of Buckling*, Ph.D. Thesis, MIT, Department of Electrical Engineering and Computer Science, (May, 1994).

[3] E. Demaine and M. Demaine and J. Mitchell, "Folding flat silhouettes and wrapping polyhedral packages", *Annual Symposium on Computational Geometry*, Miami Beach, Florida, (1999).

[4] H. Huzita and B. Scimemi, "The algebra of paper-folding", *Proceedings of the First International Meeting of Origami, Science and Technology* (Huzita ed.), Ferrara, Italy, (1989).

[5] K. Kasahara, *Origami Omnibus*, Japan Publications, Tokyo, (1999).

[6] T. Kawasaki, "The geometry of orizuru", *Proceedings of the Third International Meeting of Origami Science, Math, and Education* (Hull ed.), A K Peters, (2002).

[7] R. J. Lang, "A computational algorithm for origami design", *Proceedings of the 12th Annual ACM Symposium on Computational Geometry*, Philadelphia, PA, (1996) 98–105.

[8] P. A. Lawrence, *The Making of a Fly: the Genetics of Animal Design*, Blackwell Science Ltd, Oxford, U.K., (1992).

[9] G. Odell and G. Oster and P. Alberch and B. Burnside, "The mechanical basis of morphogenesis: 1. Epithelial folding and invagination", *Developmental Biology*, Vol. 85, (1981) 446–462.

[10] M. Resnick, *Turtles, Termites and Traffic Jams*, MIT Press, Cambridge, MA, (1994).

[11] J. Slack, *From Egg to Embryo, second edition*, Cambridge University Press, U.K., (1990).

[12] D. Arcy Thompson, *On Growth and Form, abridged edition*, Cambridge University Press, U.K., (1961).

[13] M. Yim and J. Lamping and E. Mao and J. G. Chase, "Rhombic dodecahedron shape for self-assembling robots", Xerox PARC SPL TechReport P9710777, (2000).

Part Three

Origami in Education

More and more of today's teachers are finding that origami has much to offer as an instrument for experimenting with mathematical, scientific, and educational ideas. They understand the value of having a concrete view of an abstract idea: Origami helps these teachers offer hands-on experiences to their students in order to facilitate access to abstract ideas. Fortunately, some of these teachers came together to share their findings. These proceedings of the Third Meeting on Origami Science, Mathematics, and Education show a range of the current professional uses of origami in helping students approach complex ideas.

Two papers deal with origami as a tool used to facilitate the study of English as a Second Language. Deborah Foreman-Takano explains the foundation and effects of a foreign language program she developed to promote awareness of audience with her university students in Japan. By folding origami, the students practice culturally different techniques for communicating in an unfamiliar language. This enables them to work on the content of their English expressions as well as their grammar. Lillian Yee Ho incorporates origami techniques as a mechanism for practicing vocabulary with non-English speaking adults. Her paper presents the various folded devices the students make to help each other reinforce their langugage skills.

Several papers demonstrate the marvels available for mathematics education through folding processes. Krystyna and Wojciech Burczyk approach the idea of building an unlimited variety of polyhedra from a single unit design. By analyzing the properties induced by the edge lengths or the angles of the unit, they predict the resulting polyhedra. Charlene Morrow addresses the question of ways to color the faces of a polyhedron: Is it possible to select colors so that each face meeting at a vertex has a different color? This type of question is well-studied within research mathematics, but Morrow finds a way to introduce the subject to secondary school students. Eva Knoll explores a crease pattern within a circle and discusses an educational experiment to determine its trigonometric properties. Her amazed students create a very large polyhedron which can be used to analyse the circular crease pattern. Emma Frigerio introduces a simple traditional form—the paper cup—and leads readers through the related fundamental geometric concepts. Her examples of various coasters discovered by her students by locking the cups into rings are rewarding. V'Ann Cornelius and Arnold Tubis demonstrate the making of a simple isosceles box and explain the trigonometric issues involved in generalizing the directions to apply to boxes of any depth, height, and/or apex angle.

Kazuo Haga, the keynote speaker for this section, brings joy to folders with his wonderful presentation on discovering geometry. He defines his term *origamics* and leads readers to a point where they can share his "pleasure of discovery".

Another application of origami in the world of science comes with the development of origami as an analogy. Norm Budnitz approaches the abstract concept of the genome in organism development. As a means of communicating the complexity involved, he presents a crease pattern for a simple model. The challenge is to guess what it will be, and given that information, what a similar crease pattern will become. He likens this situation to the analysis of genetic material for similar organisms.

Many people working with students of mathematics wonder what origami exercises are available in teacher text books. Jack A. Carter and Beverly J. Ferrucci did a study of ten preservice textbooks. They analyzed the origami opportunities included in the books, and present their findings here.

It is important to have an arena for educators who recognize the potential of folding paper as an educational tool. Without such an arena educators are isolated and can't benefit from the wealth of experience that others can share. Resources for origami in education are few, and too often teachers interested in origami feel they have to "invent the wheel" themselves to use origami in their classrooms. We thank those who stepped forward to share their findings and enter into the discussion of the future of origami in education.

V'Ann Cornelius

Applications of Origami to the Teaching of Sophisticated Communication Techniques

Deborah Foreman-Takano

1 Introduction

The original use of origami in formal Japanese education was in the Meiji era (1868–1912) with kindergarten and elementary school children, and according to origami historian Okamura Masao, was patterned after its use in Germany [10]. Okamura mentions that later,

> ...because origami required following precise directions, origami was not well–received during the Taishou period (1912–1926), a time when educators favored placing importance on originality and creativity. Origami was deemed as lacking in these qualities.

Now, origami is again being used in the education of young children in Japan, and has devotees of all ages there and all over the world. This is evidenced not only by a large number of books and other hardcopy publications, but by many origami-related web pages maintained by individuals and organizations (see, for example, [2], [7], [9], [14]). There are also examples of its applications to the teaching of English as a second language (ESL) and as a useful prop for practicing a variety of other skills on a simple level [3], [12].

However, the relevance of origami to the teaching of sophisticated communication techniques to adults has received no significant attention. I find origami useful when teaching Japanese university-level English classes on public and interpersonal communication. In particular, I found origami useful to:

(1) teach Japanese students about the world-wide face of origami, which helps them to think about how aspects of Japanese culture might find relevance in various non-Japanese contexts,

(2) teach them various interpersonal communication techniques, which they could hone without necessarily speaking a lot in a foreign language or using difficult vocabulary, and

(3) teach them basic public presentation techniques, including selection of topics, audience research, adaptations for audience size, and configuration and use of visual aids.

The first can be accomplished over the duration of any course using origami that includes models and written materials from various countries. One can analyze what they reveal about (a) attitudes toward things Japanese and/or (b) special adaptations made to Japanese works or ideas that reflect other cultures' customs and interests.

In this paper, however, I will focus on the second and third areas of possibility: cultivating an awareness of techniques that facilitate interpersonal communication, particularly (but not exclusively) in English, and cultivating an awareness of audience.

2   Background: The Teaching of English in Japan

Before examining how origami can be useful to help Japanese university students communicate effectively in English, it is important to understand what kind of experiences the students have previously had when learning English.

In Japan, English is taught as a foreign language, rather than as a second language. English is not required for living successfully in Japan; the stimulus to learn it may come from incentives in the work environment, but not usually from the society itself. It is different from the situation in India, for example, where English is actually a second language in which official business is conducted. Thus the question is: Why should anyone in Japan—particularly, let us say, college students majoring in a variety of areas not directly related to language—learn spoken English?

Many students have an interest in developing at least a rudimentary ability to talk with non-Japanese-speaking people. In Japan, a non-Japanese speaker is likely to be a nonnative speaker of English, probably an Asian for whom English would also be functioning as a lingua franca in the exchange. But the Japanese often do not have the confidence to use it orally, since their training is predominantly in reading and writing. Such training could be helpful, except that English is not really taught as something to communicate with (even on paper), but essentially as a "code of Japanese" to be memorized and reproduced on an exam. It is one of many subject areas in the secondary school system that is used to teach discipline to the young Japanese mind. Thus, a typical student's goal is to make as few mistakes as possible and attain good test scores. In their use of English, students are not in the position of having to consider how anyone else feels, in terms of content, understanding, or enjoyment. Understandable reluctance sets in when someone raised in the Japanese culture faces communicating with someone from a different culture in a foreign language. The preoccupation of the person acculturated to Japanese culture is: "What if I make a mistake?!" This is not meant in the sense of, "What if the person is inconvenienced by not being able to understand what I am saying?" but, "What if the person can't understand me and so has a low opinion of me?!"

2.1   Why Origami?

I started using origami at my Japanese university in spoken English classes to provide a distraction from this preoccupation. The tactic in itself is not very unusual; most teachers accustomed to participatory classrooms find that they need a trick of some kind to get students in Japan to be less focused on "perfect production" and more focused on, simply, production.

Furthermore, one of the age-old problems foreign language teachers face is making class practice directly applicable to real-world experience. This is particularly true in Japan, where foreign languages don't have general applicability to everyday life. Using origami creates a pleasant, nonthreatening, and independent context that can be duplicated in various environments outside of the classroom. The amount of language used can vary naturally. The vocabulary need not be difficult, but is nevertheless appropriate for dialogues between adults. Perhaps most importantly, the student/folder can, with a minimum of effort, maintain a high degree of control over these elements.

2.2   Interpersonal Communication Hurdles

There are a number of components to the process of social communication (see [1] for a concise treatment) which can present obstacles to the accurate conveying and interpretation of messages. What creates such obstacles is the lack of awareness that people have in the communication process.

Many textbooks dealing with the teaching of oral English place emphasis on vocabulary (including idioms) and syntax, using artificial "dialogues" that ignore the between-the-lines pragmatics that are pivotal to conversations in real life. While this approach is compatible with the sort of rote-memorization, one-right-answer methodology Japanese students are used to through high school, it does not provide any guide to understanding many of the obstacles to effective communication. Those texts that do attempt to deal with the conversation skill itself tend to focus on "gambits"—simple tactics presented in simple settings which ignore two considerations: (1) real-life situations are rarely as cut-and-dried as those presented in textbook dialogues; and (2) there is little help provided for the students to apply the highly artificial role playing they do in the classroom (e.g., two Japanese students acting the roles of a father and son speaking English to each other) to their lives outside of the classroom.

For an alternative example, Reinhart and Fisher [11] have developed a text-book which does deal with some of these important conversational/social dynamics. However, this book is not usable with elementary- to low intermediate-level English students; the vocabulary necessary in addressing these complex-ities looms as at least as big an obstacle as the pragmatic issues themselves.

I began thinking about origami as a solution to these educational problems after learning that a book I had planned to use for a course was out of print. It occurred to me that an origami book might turn out to be an intriguing resource; it was certainly the last kind of book the students would expect to be using. I considered the basic characteristics such an origami/text book would need. As I commenced studying origami, for the first time since my elementary school days, I realized more and more how appropriate it might turn out to be for awakening students to various communication techniques without other complicating factors coming into play.

Japanese students enter university having had six years of English instruc-tion, and therefore have a large amount of stored material on which to draw. What they have not had is training in how and when to use all this stored vo-cabulary and grammar information, particularly in oral-aural modes. I thought it would be efficient to get them to pull out all this information, dust it off, and learn how to put it to friendly and practical use.

3   Starting a Conversation: Sample Dialogues

The idea of using origami to facilitate conversations is introduced to students in the following way:

It is not easy be begin a conversation with someone you don't know. You can start with the usual questions about the person's name, where she or he is from, and perhaps questions about hobbies or interests. This kind of beginning is usually okay, but there could be several problems:

(1) The answers to the questions might use difficult or colloquial vocabulary. Each person has her or his own speech habits. When you meet someone for the first time, you have not yet had a chance to get used to the accent or the particular speech habits of that person. Moreover, you are likely to be a little bit nervous talking to the person for the first time: This increases your difficulty in understanding.

(2) The person is likely to ask you the same questions that you asked her (or him). This will require you to explain your hobbies and other things about yourself. While you may be able to do this in a simple way, your answers may lead to more complicated questions and comments from your conversation partner.

(3) It is also possible that the person may be tired of these questions and topics. After all, they are very usual and typical questions that are asked between strangers talking for the first time. If you do the same thing, you will not make yourself memorable.

(4) Perhaps you don't want to begin a conversation by discussing personal details. This can often be a good and safe decision. However, you would like to have a friendly introductory exchange with people that you meet.

3.1   Origami Is a Good Start

Although many people (including many Japanese) do not know it, origami is folded all over the world. There is origami for children and origami for adults. There is simple origami, as well as, various levels of complex origami. Each has its own charm. People who are unfamiliar with origami are curious about it; people who are familiar with origami enjoy doing it and talking about it. Everyone considers origami a part of Japanese culture, so your folding will appear quite natural. If you are among strangers and begin folding origami, you have several advantages:

(1) You will not look the same as everyone else in the room. You will look interesting. People who do not know origami will wonder what you are doing, and people who know origami will wonder what model you are folding. People who are curious, and people who are interested in you, will have an easy way to start talking with you: They will ask you about your activity.

(2) Talking about origami does not require complicated English. You can explain folding with a combination of words and gestures. You can use many words, or you can use few words, whatever suits your mood. Teaching simple folding is not difficult, but produces some beautiful results. For many people it will be a unique introduction to Japanese culture, and it will be easy to talk for a while with these people.

(3) Discussing origami is not offensive. You will not be required to talk about politics or religion or other topics that can make people emotional and can make communication difficult.

3.2   Teaching Materials

Following are some dialogues I have made as illustrations of how to deal with several kinds of situations and people. There is no dearth of "sample dialogues" in books aiming to teach spoken communication. But, as mentioned before, they are seldom representative of language/turn-taking that occurs in the real world, and ignore all but the simplest pragmatics issues. These are not necessarily difficult issues to get across, as the material below shows, and they use vocabulary and syntax already familiar to the students. Each of the four dialogues presented here addresses one or several common and real communication dynamics/problems. The students seem surprised to learn that these things can actually occur in English conversations, although they recognize that they happen in Japanese conversations; English thus far for them has been, as pointed out previously, a largely artificial and formulaic exercise.

The main point of all the conversations below is to introduce ways to make the conversation "flow" in each case, and at the same time, to show how to give evidence that close attention is being paid to the conversation partner. The concept of flow is given short shrift in English courses at the secondary level. Even with long readings, the methodology used is what is generally referred to as "grammar-translation," involving the rendering of very small chunks of English text individually into an unnatural "translation Japanese," with little attention to nuance or context. It is understandable that this deficiency could carry over into the spoken-language domain.

Possible locations for the following situations—familiar places for origami activity to all paperfolders—are listed for the students:

(1) In a train, or plane, or other public transportation.

(2) In the waiting room of a doctor's office or a hospital.

(3) On a bench at a bus stop, train platform, or airport waiting area.

(4) In the school cafeteria.

(5) In the library.

(6) In a hotel lobby.

(7) Waiting in line for something.

(8) Waiting for a club meeting to start.

(9) Waiting for a meal in a restaurant.

I

B: May I ask what you are making there?
A: Oh, I'm making an origami elephant.
B: What kind of elephant?
A: Origami. It's Japanese paper folding.
B: It looks hard to do.
A: This elephant is a little tricky. Some things are easier, though.
B: Really? ...It's fascinating.
A: I have some extra paper. Would you like to try something?
B: Oh, would that be all right? Thank you.
A: Here—what color paper would you like?
B: Look at all those colors! What are we going to make?
A: Hmm ... How about a (balloon)?
B: A (balloon)? Out of paper?
A: Yeah. It's easy.
B: OK. Then, let's see—I'll take (yellow).
A: OK. Just pull it out of the package ... I'll take (red).

Situation I addresses several points. The first is the concept of using long forms instead of short forms to increase the politeness of utterances, something that is common in the Japanese language:

(1) "May I ask (what) ..." instead of just "(What) ..."

(2) "Would you like to try ..." instead of a simple "Will you try ..."

(3) "Oh, would that be all right? Thank you," instead of just "Yes, thank you."

The second concept is that of showing modesty in English. Apart from any perception the students might have that this would be difficult, there is a popular stereotype of Americans as being "frank" and unhesitatingly indicating pride in what they can do. Japanese are raised to avoid doing this, and there are numerous Japanese linguistic conventions which are designed to convey humility. I like to introduce the students to ways that Americans show humility, and here I use the phrase "a little tricky" as a humble characterization of something difficult that one has accomplished.

Third, the flow of the conversation—which has the aim of making the dialogue a pleasant exchange for both parties rather than something resembling a police inquisition—is facilitated with such phrases as "I have some extra paper," "Here—", "Just pull it out of the package," and pause markers such as "Hmm" and "let's see..."

II

(A notices B watching the folding.)

A (Smiles at B): Do you like origami?

B: Yes, I learned it when I was a kid. I don't think I can remember how to do it, though.

A: What did you make?

B: I used to make a crane ... and a fox puppet ... and some other things.

A: I've got some more paper here. Would you like to see how much you can remember?

B: Yeah, that might be fun! Thank you.

A: Just pick your favorite color out of the package there.

Situation II focuses on ways to indicate active listening in a conversation. First of all, A responds to the attention exhibited by B in the origami folding. Secondly, A asks a question that follows logically from what B says, indicates personal interest, and elicits more than just a short answer from B. Third, A further proves how well s/he has been listening by saying, "Would you like to see how much you can remember?" instead of just something unconnected to the previous content, like, "Will you try?"

III

(B is sitting in a chair, doing nothing.)

A: (Folding origami) Are you waiting for someone?

B: (Nods.)
A: Would you like to try folding some paper?
B: No, thank you.
A: (Smiles) Oh, OK.

Situation III addresses the issue of rejection of a conversation attempt. In my thirty years in Japan, I have found almost nothing in textbooks to suggest that someone's attempt to talk with someone else might be met with silence or surliness; most "sample dialogues" show people indicating politeness, kindness, and delight, no matter what the topic is. This is obviously the attitude everyone is hoping for when they strike up a conversation with someone nearby, but let's face it: Bad moods and bad circumstances exist. Students should be at least intellectually prepared for the fact that someone might be unwilling to talk with them, and that this is not likely to be their fault. Situation III gives them a chance to work on intonation and a facial expression that will allow them to exit the situation gracefully and with pride intact.

IV

(B is watching A folding)
B: Hey, that's origami, isn't it!
A: Yeah. You recognized it!
B: Are you Japanese?
A: Yeah.
B: All Japanese know origami, right?
A: Well, we learn some simple things when we are in elementary school. Many Japanese forget how to fold, though.
B: But you remembered!
A: Some of us continue. There are many origami clubs in Japan, but many Japanese don't know about them. There are many kinds of folding—difficult folding, too.
B: What you're making looks difficult!
A: It's a little tricky. What country are you from?
B: The U.S.
A: There are origami clubs there, you know. There are even some origami conventions!

Situation IV is designed to teach the students ways to humor people who might be behaving boorishly. The concept of humoring someone does, I believe, allow dignity to prevail, and allows at least the boor to enjoy the exchange.

Here, B seems to be prepared to have all her/his stereotypes confirmed by
trotting out a series of yes/no questions. But A's comments and responses are
examples of long and short answers, yielding control and taking control, that
can be employed in many conversations.

4   The Concept and Relevance of Audience

As I considered how origami was working in my "English-Speaking Cultures"
classes, I saw it could be applied to the development of speech and presentation
expertise.

A typical project in these speech classes is, of course, the actual preparation
and delivery of a presentation. Public speaking textbooks (see [4], [5], [6],
[8], [13]) give good, organized advice on the procedures themselves. But the
student actually spends a great deal of time worrying about the choice of topic
and audience research. Indeed, students with little public speaking experience
can be distracted by choosing and organizing a topic and spend little time on
integrating it into the speechmaking process.

I tried using origami in this kind of course for several reasons:

(1) Students in the class were usually anticipating study abroad or extensive
    use of English in their future jobs in Japan. They were surprised how
    little they knew about origami, which they considered part of Japanese
    culture, and were interested in investigating it.

(2) They could work together on the research of the various aspects of
    origami, including following up with investigations of what the pref-
    erences of the class/audience would be with respect to type of model to
    be tried.

(3) They would be forced to rehearse their presentations many times (a step
    often skipped by busy and inexperienced speech students cutting corners),
    in order to test the timing and clarity of the folding explanations.

(4) Their presentation would be getting immediate feedback in terms of how
    successful their classmates would be at folding the models.

(5) This would help them recognize the responsibility of the speaker to make
    sure the audience understands the presentation. (The Japanese often
    consider the responsibility for understanding to rest with the listener.)

I found that the use of origami did reduce complications associated with the
preparation of speech material. The immediate feedback inherent in this kind

of presentation allowed the student presenters to focus accurately on where improvements needed to be made for the next project. Also, time constraints imposed on the presentations required the presenters to be creative in handling the choice of model and how it was taught to the class: Some groups assigned certain origami bases as homework so the in-class folding would take less time; some groups had each of the three members teach one-third of the class. The solutions and attempts at solutions were creative.

The audience-awareness element in a public speaking exercise, like the conversation partner element in a dialogue, takes oral language out of the realm of one-way communication. Consciousness of these elements allows students to put classroom study into a kind of real-world perspective, considering not only their own role, but the role of others in a communicative act.

References

[1]   Benjamin, J. B. *Communication: Concepts and Contexts*. New York: Harper & Row Publishers, Inc., 1986.

[2]   Budai, P. Origami Links, 2002.
        Available: http://peterbudai.tripod.com/Origami/Orilinks_en.htm

[3]   Cornelius, V. (Ed.). *Proceedings of the First International Conference on Origami in Education and Therapy*, OrigamiUSA, 1995.

[4]   Fisher, T. and T. L. Smith., *Icebreaker: A Manual for Public Speaking*, Prospect Heights: Waveland Press, 1985.

[5]   Gronbeck, B. E., R. E. McKerrow, D. Ehninger, and A.H. Monroe. *Principles and Types of Speech Communication [Eleventh Edition]*, Glenview: Scott Foresman/ Little, Brown Higher Education, 1990.

[6]   Hanna, M. S. and J. W. Gibson. *Public Speaking for Personal Success*, Madison: Brown & Benchmark Publishers, 1987.

[7]   Hull, T. Origami-Math Bibliography, 2002.
        Available: http://web.merrimack.edu/~thull/oribib.html

[8]   Klopf, D. W. and R. E. Cambra, *Personal and Public Speaking (3rd Ed.)*, Englewood: Morton Publishing Company, 1989.

[9]   O'Hanlon, S. Dr. Stephen O'Hanlon's Origami Page, 2002.
        Available: http://www.paperfolder.org/

[10]   Okamura, M. "The History of Origami," 1999.
        Available: http://www.origami.gr.jp/People/OKMR_/history-e.html

[11]   Reinhart, S. M. and I. Fisher. *Speaking and Social Interaction (2nd Ed.)*, Ann Arbor: The University of Michigan Press, 2000.

[12]   Smith, J. (Ed.). *Proceedings of the Second International Conference on Origami in Education and Therapy*, The British Origami Society, 1991.

[13]   Verderber, R. F. *The Challenge of Effective Speaking*, Belmont: Wadsworth Publishing Company, 1988.

[14]   Wu, J. Joseph Wu's Origami Page, 2002. Available: http://www.origami.as.

Origami and the Adult ESL Learner

Lillian Yee Ho

1 Introduction

Origami has long been used in elementary and secondary education. It relates to many different curricula including mathematics, science, social science, art, and language arts, as well as in special education and therapy (see [2]). However, little has been written in professional journals on using origami in teaching ESL, particularly in the adult setting.

1.1 The Rationale in ESL Classroom Instruction

Drawing, music, and poetry has played an established role in the adult ESL classroom. By the same token, origami is still another medium to add to the adult teacher's bag of tools. It can be incorporated into lessons teaching basic ESL skills of listening, speaking, reading and writing. In addition, it lends itself to other areas of study such as multicultural awareness, ecology, and art appreciation. Adopting origami in an ESL lesson will open the class to a variety of perceptual learning modes; namely, tactile and kinesthetic as well as visual and auditory. In terms of social skills, learning to fold can encourage cooperative learning and students teaching each other.

1.2 The Problems

While there many of origami books on the market, the materials do not always lend themselves readily to the adult ESL classroom. The subject matter may be inappropriate (e.g., animals, toys, or dinosaurs). The models may be either too complex or too juvenile. The directions, written for native English speakers, are often too technical for the average ESL class. However, it is my contention that origami diagrams and instructions, like any other authentic materials can, and should, be carefully chosen and adapted for adult ESL.

Materials aside, ESL teachers may be reluctant to use origami because they a) do not see a connection, b) cannot fit origami into the curriculum goals and, c) find the preparation and/or class activity too time consuming. Here again, understanding origami's relationship to ESL and careful planning is key, which brings us to the raison d'être of this paper.

1.3 Thesis and Goals

The case of whether using origami in the adult ESL classroom is pedagogically sound is a question worthy of serious exploration. The purpose of this paper is to introduce origami as a valid educational tool for ESL teachers and others in the field. There are many factors to consider when developing one's own ESL lessons. Tips on how to use origami successfully in class and some practical folding ideas and models to use with various levels of adult students will follow. Suggestions on how to incorporate origami in ESL will be offered as well. Hopefully other educational professionals will be encouraged to use this information and explore the topic further.

2 Developing ESL Lessons Using Origami

Given the dearth of materials for teaching origami to the adult ESL class, a teacher will inevitably need to develop his or her own. Factors one needs to consider to do this effectively include: curriculum needs (reading, writing, listening and speaking, lifeskills, competency-based tests); student background (educational experiences, culture, age, interests, diversity); learning styles (visual, auditory, kinesthetic, and social); and ongoing developments in the field (cooperative learning, task-based learning, and SCANS, the Secretary of Labor's Commission on Achieving Necessary Skills).

All the above elements should be considered foremost in developing lessons. In fact, origami will, in most cases, be relegated to a secondary role in the process. Nevertheless, if these factors are ignored, any use of origami in the classroom would be merely a recreational exercise.

When writing your own lessons, where does origami fit in? Here are several suggestions.

A. Selected models taught as an enhancement or reinforcement of content curricula such as holidays, ecology, or art.

B. Folding used as a tool for developing visuals or worksheets.

C. Storigami—folding a series of models during storytelling.

D. Models used as devices for doing grammar exercises.

E. Folding tasks specially designed to teach a language skill or competency.

3 The Fortune Teller: A Toy for the Adult ESL Classroom

If you think your adult students are too old for toys, take a shot at these two (but please don't call them that in class!). The Fortune Teller is a classic with many uses. When adapted in an ESL class the Fortune Teller truly adds a kinesthetic dimension to speaking exercises. There are two components to consider in the planning of lessons using such manipulative props in class: a) teaching the folding and b) teaching its use to practice the targeted language. When teaching the folding component, consider teaching language skills and learning strategies such as diagram reading, note-taking, learning from other students, clarification language, and problem-solving.

The Fortune Teller, also known as "cootie catcher" or "salt cellar," is a playground fold that has been popular with children for generations. This device holds many possibilities for the adult ESL learner if you adopt suitable themes. You can practice sentence construction by stringing together parts of sentences written on flaps that you open out in sequence: Manipulating the object adds a bit of surprise element, like picking a card from a deck.

To construct: The Fortune Teller is a classic origami model and can be found in many beginner-level origami books (one makes it by *blintzing*, or folding the corners of a square to the center, twice in opposite directions.) For example, see the excellent reference book *Complete Origami* by Eric Kenneway [4].

One can manipulate the finished model by holding the ridges underneath the petals with the thumbs and forefinger of both hands to open sideways, close, open longways, and close. Repeat as desired.

Constructing an ESL activity for your Fortune Teller: The fortune teller has three components, or "tiers" as we'll call them, where information can be written (see Figure 1).

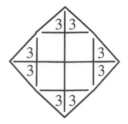

Figure 1. The Fortune Teller with the first, second, and third "tiers" shown.

1. The first tier consists of the outside petals and is the most visible. There are four petals.

2. The second tier can be viewed by looking into the model as it is held open longways or sideways. You can see four choices at a time.

3. The third tier comes into view when you open the flap underneath the second tier. The information entries tie in closely to the ones on the second tier.

The easiest way to create an ESL activity is to first have a sentence in mind that has three parts to it. For example, number, color, and clothing item would be the parts to "Do you have three red dresses?" (Actual activities will be explained in more detail below.) To prepare, make a complete Fortune Teller from an 8 1/2 inch square of paper. Write in the entries on the appropriate tiers, e.g. numbers 1–4 on tier 1; eight color names on tier 2; and eight clothing items on tier 3. Then unfold and photocopy a class set on 8 1/2 × 11 inch paper. Trim it down to a square on a paper cutter. Scratch paper that has been printed on one side can be used. Bring all this to class and teach students to assemble the model and do the activity.

Example activity: "Do you have three red dresses?" Write on the three tiers of the Fortune Teller as follows:

	Tier 1	*Tier 2*	*Tier 3*
	number	color	clothing item

Procedure:

1. Student A tells Student B to pick a number.

2. Then he/she spells the number while manipulating the teller, e.g., "two" – "t" = longways, "w" = sideways, "o" = longways.

3. Student B chooses a color, e.g., green.

4. Student A opens the flap under "green" and sees an item of clothing, e.g., pants. She asks Student B a question using all three entries. "Do you have one pair of green pants?"

5. Student B gives a short answer. After several turns, have the student pairs switch roles.

Variation: wh-question/adv/embedded clause practice:

	Tier 1	Tier 2	Tier 3
	wh-word	verb	adverb

Sample sentence: How do you speak clearly?
                  1   2   3

In this variation, wh-words, verbs, and adverbs are writen on the Fortune Teller. Student A constructs a question by manipulating the model. Student B answers with "I don't know _____", e.g., "I don't know how to speak clearly."

4   An Envelope: A Lesson for Diagram Reading

Reading diagrams and directions are life skills often taught in intermediate ESL classes. They are also essential skills used to decipher material in origami literature. Here is a lesson capitalizing on another one of origami's strengths to use in ESL.

1. After a letter writing lesson, tell class they are going to fold an envelope to enclose their letter. I recommend Gay Merrill Gross and John Cunliffe's envelope in *Easy Origami* by Gross and Weintraub [3] (which is an excellent resource in its own right for origami education ideas). Show the students the finished model.

2. Pass out some scratch 8 1/2 × 11 paper.

3. Demonstrate folding the model step by step with the class following. Encourage students to help each other along the way.

4. After the class completes the model have them unfold it.

5. Project the diagrams, without the text, on an overhead.

6. Have students reassemble their model while you read the directions and point to the steps. Instruct students not to work ahead, but to stay at

the step you are pointing to. Tell them to concentrate on listening to the directions and understanding the diagrams.

7. As you go through the directions, discuss the diagram; e.g., "Do I fold the long edges together or the short edges? Are these the long edges? What does 'crease' mean? Do I unfold or leave paper folded?" These are some of the questions that you might use to familiarize students with the language contained in the first step.

8. Hand students a worksheet with two parts. Part A consists of the original diagrams with the directions deleted. Part B consists of the directions re-typed and lettered in a scrambled order.

9. Tell students to match each of the instructions to the correct diagrams by writing its letter over the diagram.

10. When the class finishes the exercise, correct it by reading the directions in the proper order. Have the class tell you their answers.

11. Have the students copy the directions under the diagram.

12. If time permits, have the students make another model in class following the captioned diagrams. Otherwise, assign this for homework.

5 Nguyen's Many Hats

This example exercise is a storigami lesson adapted to ESL by the author. It uses an origami hat, which turns into various shapes that, when folded, serve as visual aids for a story. One tells the story while folding and manipulating the hat. Diagrams and the original version of the story can be found in the excellent origami book *The New Origami*, by Steve and Megumi Biddle [1].

Level: Beginning-High

Objective: Listening and retelling a story in the context of job occupations.

Materials: Several sheets of newsprint, some prefolded into the captain's hat if desired, and folding instructions for the hat [1].

Story: Nguyen is a young man from Vietnam. He lived by the sea. When he was a young child he would sit by the sea and dream about the future. What will he be when he grows up? Maybe he will be a captain and sail to places faraway. Or maybe he'll be a fireman. That'll be nice. He can put out fires. Or maybe he'll be an explorer. He loves discovering new places. The only problem is, there are pirates in the ocean. He heard about how pirates stole people's valuable gold and sometimes even tipped over their boats. The people were left only with the clothes on their backs.

Finally one day, Nguyen grew up. He had to put away childish dreams about being a sea captain and such. So he joined the army. After a couple of years in the army, the only job he could find was a painter, but he didn't want to do that all of his life. He would feel like such a fool. But, all the good things in life came with hard work. He knew what he had to do. He would have to go back to school and take ESL classes, and he did. He even graduated from college.

How about you? What do you want to be? Your life is like this plain sheet of paper. You can make your own hats!

Procedure: To prepare students for the listening exercise, write the occupation words on the board in scrambled order and explain them. Tell students they will be listening for these new words in the exercise.

1. First listening: Instruct students to listen for the listed words and raise their hands when they hear one. Meanwhile, the teacher tells the above story in its entirety, while folding the piece of newspaper into hats corresponding to the story. Note: You will need to start out with another captain's hat in the second part of the story because the first hat will be torn for the shirt.

2. Second listening: Tell the story again, but this time stop mid-way through and ask some yes/no questions. Finish telling the second half and repeat the procedure.

3. Do T/F questions about entire story. Alternatively, you can make false statements and ask students to tell why the statements are false.

4. Ask wh-questions: e.g., "What did Nguyen want to be when he was little?"

5. Do the entire folding sequence again, this time with the class retelling the story.

6 Origami ESL Exercises with a Cooperative Learning Focus

Using language is by nature a communicative activity. When there is a task to be accomplished, learners must communicate with each other to complete it successfully. This is called cooperative learning in the educational field. The following exercises are ways you might incorporate this aspect of language teaching into a lesson while using origami as the subject of the task.

General procedure:

A. Choose a modular model with pieces that require several steps. The model itself should not require too many pieces, perhaps 12 or less.

B. Have the class pair up into A/B partners. All the As leave the room while you, the teacher (hereafter known as T), show the Bs how to fold the first half of the module.

C. The As return to the room and each rejoins his or her partner B. B shows A how to fold the module up to the point where T stopped.

D. Then the Bs leave the room while T shows the As the remaining steps to complete the module.

E. The Bs return to the room and rejoin their partners. Then A shows B how to complete the module.

F. Each A/B pair must work together to make the required number of modular pieces to complete the entire model.

G. T then provides diagrams for each pair on how to assemble the pieces.

H. The faster students are encouraged to help the slower ones with the diagrams.

I. Follow up: A worksheet with separate columns of diagrams and instructions written in scrambled order could be given to students to match up for homework.

Variation 1: Instead of choosing a modular model with similar pieces, use a composite model with 2 different pieces such as an animal with a separate head and body, a clown face with a hat, or a flower with a stem and leaf. Stop with step E. Alternatively, combine it with a greeting card writing exercise and use the origami to decorate the student made card.

Variation 2: For lower levels, type out a cloze exercise (leave out blanks for students to fill in) out of the oral instructions you will give to each half of the class. Do the cloze exercise by dictating the appropriate part of the instructions for steps B and D first. Then demonstrate the folding, repeating those instructions and allow the students to read them to their partners as they demonstrate the folds in steps C and E.

Variation 3: This is an adaptation of the classic jigsaw exercise commonly used for ESL reading or listening activities. It satisfies several of the SCANS objectives, such as students acquiring critical thinking skills, organizing information, and teaching each other.

For this exercise use a modular model with at least eight pieces, no more than 24. In step F, put two pairs of students together to make a group of four. After most of the groups have completed most of the pieces, have each group send a rep to join the expert group. While the others are finishing up on the pieces, T will show the expert group how to assemble the model. The experts will return to their home groups and explain the assembly process to their members.

Tips:

1. Practically all of language is delivered through the oral channel, so to be effective, make sure the vocabulary is clear. Preview with the entire class terms like crease, fold, corner, diagonal, piece, etc., if necessary. Also, use language that is not too difficult for the level of your students.

2. Repeat instructions as necessary and unfold model after steps B and D. Have students give you the step by step directions as you refold.

3. Remind class that this is, after all, an ESL class and they must use English as much as possible when explaining directions to their classmates.

7   Origami for ESL Students in Content–Based Classes

So far, the lessons set forth in this paper have been geared toward a general ESL adult school population. They may be less suited for students in a more rigorous academic ESL program.

Some colleges have ESL programs that prepare their students to enter an academic field of study. Often the student is concurrently enrolled in such courses while honing their English language skills. Here are some ideas on how origami might be used in these situations:

Academic style readings: Teachers can take an article on origami with a scholarly focus in history, math, or science and develop comprehension questions, vocabulary and other exercises.

Student academic research topics: Students can write a paper on questions such as the following: How has origami influenced the culture of Japan?

What are some practical applications of origami in geometry? Is origami an effective three-dimensional visualization tool for molecular structures?

A student presentation topic: Origami can be one in a list of several "how to" topics that students can choose from to prepare a timed oral report.

Creating of hands-on origami models for an ESL bridge class to a content area classroom: Students can be given an assignment to create and report on diagrams and text for a computer class, structural forms for architecture, or design models for graphic or industrial design.

Experimentation with origami use in therapy: Students in clinical or physical therapy fields can report their findings in the ESL session.

8 Summary

It is not only possible, but can be beneficial to use origami in an adult ESL classroom. Origami can be relevant to adult interests and a viable teaching tool. A teacher must first consider factors such as curriculum, student needs and his or her own experience in developing and using ESL activities. Combine these with origami and the result will be guaranteed success. This paper is only a beginning. It is hoped that other ESL teachers will experiment with the ideas presented here and explore other possibilities.

Acknowledgements

Special thanks to my colleagues in the ESL Department at the City College of San Francisco: Eunice Lew, my co-presenter in the original workshop on this topic for her inspiration and tireless support and Patricia Seid; for helping with the layout and revision of this paper.

References

[1]   Steve Biddle and Megumi Biddle. "The Captain's Hat Story," in *The New Origami*, St. Martin's Press, New York, (1993) 62–67.

[2]   *COET '95 Conference on Origami in Education and Therapy*, Origami U.S.A., New York (1995).

[3]   Gross, G. M. and T. Weintraub. *Easy Origami: step-by-step projects that teach across the curriculum*, Scholastic Professional Books, 1995.

[4]   Kenneway, E. *Complete Origami*, St. Martin's Press, New York, (1987).

[4]   Pearl, B. *Math in Motion*, Origami in the Classroom (1994).

Exploring the Possibilities of a Module

Krystyna Burczyk and Wojciech Burczyk

1 Our Goal and Tools—The Definition of a Problem

Rona Gurkewitz ([4], [8]) defined a modular system as a family of polyhedra models made from the same module. We ask if all regular (Platonic and Archimedean) polyhedra form a family for a specific edge module. To answer this question we face the problem: Is it possible to assemble all regular polyhedra using only one type of edge module and how do we do this?

This question, and the study it requires, would serve as a great opportunity for students to practice their geometry and reasoning skills in solving this complicated problem. The methods we present can serve as a guide for educators in leading students on such research projects.

2 Module Types

Modules used for polyhedra building can be classified into three categories: edge modules, vertex modules, and face modules.

In the case of *edge modules*, each edge of a polyhedron corresponds with a module and modules join in a vertex. Tomoko Fuse's open frame 2 module [2], Pietro Macchi's dodecahedron module [6], Francis Ow's fullerene module [1], and David Mitchell's outline dodecahedron [7] are the examples of such modules.

In the case of *vertex modules*, each vertex of a polyhedron corresponds with a module and modules join across an edge. Lewis Simon's gyroscope module and triangular gyroscope module [8] and Rona Gurkewitz's truncated gyroscope module [8] are the examples of such modules.

In the case of *face modules*, each face of a polyhedron corresponds with a module and modules join along an edge. Nick Robinson's rhombic dodecahedron module [7] is the example of such a module.

3   An Edge Module

This often, used edge module consists of two isosceles trapezoids connected along the longer side. Each trapezoid has a flap on one side and a pocket on the other side. The trapezoids' longer base forms an edge of a polyhedra. The *angle of a module* β is a characteristic of the module. This angle is defined as an angle between the line which becomes the polyhedra edge and the edge of a pocket (the acute angle of the trapezoid). We call such module a β-module.

We can construct any acute angle module (more precisely, we can approach with arbitrary precision) if we use the following construction:

1. Divide a side of a piece of paper into 2^n equal pieces (by folding it in half). Fold a corner to one of the lines, where the crease begins at the other corner or at one of the division points.

2. Bisect (one or several times) the constructed angle.

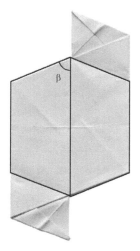

Figure 1. An edge module.

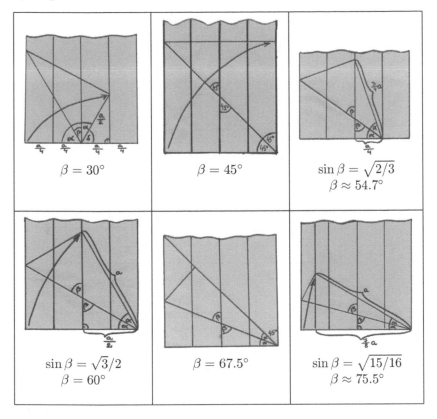

Table 1. Constructing angles for various edge modules.

Table 1 shows examples of edge modules constructed with such a technique.

Many different modules can be constructed in a similar way. Such modules behave similarly to the modules shown above.

3.1   The Opening Angle

The *opening angle* α is the maximum allowed angle between two adjacent modules.

The opening angle is the double of the angle of the module: $\alpha = 2\beta$.

It is easy to determine if a given polygon can be assembled with a given module. The opening angle of a module (twice the module angle) must be greater than or equal to the interior angle of a polygon. Table 2 shows combinations of polygons and modules satisfying this condition.

Figure 3 shows such polygons made with a 67.5°-module.

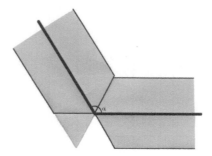

Figure 2. α is the openng angle.

Module angle	30°	45°	54.7°	60°	67.5°	75.5°
3-gon (triangle)	✓	✓	✓	✓	✓	✓
4-gon (square)		✓	✓	✓	✓	✓
5-gon			✓	✓	✓	✓
6-gon				✓	✓	✓
8-gon					✓	✓
10-gon						✓
12-gon						✓

Table 2. Which module angles will work for which polygons.

3.2 The Sinking Angle

The above reasoning gives us an easy criterion for when a polyhedron can not be built with a specified module. But comparison of the opening angle of a module and the interior angle of a polygon is not sufficient. When we assemble a polyhedron model with edge modules, it is necessary to have enough space to put all the trapezoids inside. Greater module angles give us more polygons, but the same angle limits the possibility of joining the polyhedron's edges.

Figure 3. Polygons made from a 67.5°-module.

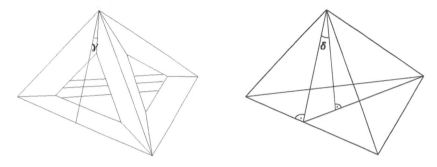

Figure 4. Illustrations of the sinking angle γ and maximum sinking angle δ.

When edge modules form a polygon face, the trapezoid sides of the modules sink inside the face. There is an edge where two module sides meet. The angle between this edge and a polygon (polyhedron face) is called the *sinking angle* (γ). (See Figure 4.)

The *maximum sinking angle* δ is the angle between the face and an "altitude line" drawn from a vertex to the center of the polyhedron. This will be the largest sinking angle that the polyhedron will allow. Thus the condition $\gamma \leq \delta$ must be satisfied if the modules are to form a vertex in the polyhedron model.

If we connect modules to form a polygon and define the angles as follows:

   β is the module angle,

   γ is the sinking angle, and

   θ is half the polygon interior angle, i.e., $\theta = (n - 2)/n \times 180°/2$

   (n is the number of polygon sides),

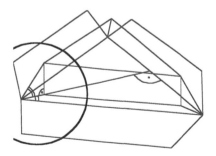
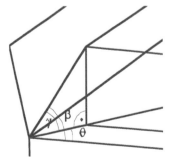

Figure 5. Comparing the angles β, γ, and θ ($\bullet$ denotes right angles).

Module angle	30°	45°	54.7°	60°	67.5°	75.5°
3-gon (triangle)	0°	35.26°	48.14°	54.74°	63.78°	73.20°
4-gon (square)		0°	35.19°	45°	57.2°	69.26°
5-gon			10.55°	31.72°	49.4°	64.79°
6-gon				0°	40.1°	59.95°
8-gon					0°	49.14°
10-gon						35.88°

Table 3. Approximate sinking angles for various polygons.

then the following relationship is satisfied:

$$\cos\gamma = \frac{\cos\beta}{\cos\theta} = \frac{\cos\beta}{\cos\left(\frac{(n-2)}{n} \times \frac{180°}{2}\right)}.$$

Table 3 shows sinking angles for polygon-module combinations. (All values are approximations, of course.)

We can consider the edges of a polyhedron meeting at one vertex as forming a pyramid. When we project these edges onto the base of the pyramid, we get the following situation (only two edges and their projections are shown):

δ is the maximum sinking angle,

θ is the half of the polygon (face) interior

angle, i.e., $\theta = (n-2)/n \times 180°/2$

n is the number of the polygon's sides),

ϕ and is the half of the base central

angle (projection of θ).

Then the following condition is satisfied: $\sin\delta = \tan\theta/\tan\phi$.

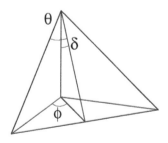

Figure 6. Comparing δ, θ, and ϕ.

	$\sin \delta$	maximum sinking angle δ	face
Tetrahedron	1/3	$\approx 19.47°$	triangle
Cube	$1/\sqrt{3}$	$\approx 35.26°$	square
Octahedron	$1/\sqrt{3}$	$\approx 35.26°$	triangle
Dodecahedron	≈ 0.7946	$\approx 52.62°$	pentagon
Icosahedron	≈ 0.7946	$\approx 52.62°$	triangle
Cubooctahedron	$\sqrt{2}/3$	$\approx 54.74°$	triangle
	$1/\sqrt{2}$	45°	square
Icosidodecahedron	≈ 0.9342	$\approx 69.09°$	triangle
	≈ 0.8507	$\approx 58.28°$	pentagon

Table 4. Approximating maximum sinking angles for various polyhedra.

If our polyhedron is regular, then $\phi = (360°/m)/2$, where m is the degree of the vertex. In such a case, the maximum sinking angle is the same for each face.

In the case of the cubooctahedron and the icosidodecahedron, we get two maximum sinking angles—one for triangles and the second for squares or pentagons. Here $\tan \phi_1 = \sin \theta_2 / \sin \theta_1$, where θ_1 and θ_2 are angles of two faces in one vertex. Table 4 shows our results.

4   Polyhedral Families for Edge Modules

When we combine the opening angle condition and the sinking angle condition, we get a necessary and sufficient condition for the possibility of assembling a polyhedron model with edge modules. Table 5 shows the results.

In the following parts of this article, we describe three techniques that allow us to go beyond the limitations shown above. We will use triangulation, domes, and reversal of a module.

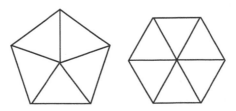

Figure 7. Pentagon and hexagon spokes.

Module angle	30°	45°	54.7°	60°	67.5°	75.5°
Tetrahedron	✓					
Cube		✓	✓			
Octahedron	✓	✓				
Dodecahedron			✓	✓	✓	✓
Icosahedron	✓	✓	✓	✓	✓	✓
Truncated tetrahedron				✓		
Truncated cube					✓	?
Cuboctahedron		✓	✓	✓		
Truncated octahedron				✓	?	?
Snub cube		✓	✓	✓	✓	?
Lesser rhombicuboctahedron		✓	?	?	?	?
Greater rhombicuboctahedron					?	?
Truncated dodecahedron						?
Icosidodecahedron		✓	✓	✓		
Truncated icosahedron				✓	✓	?
Snub dodecahedron		✓	✓	✓	✓	
Lesser rhombicosidodecahedron		?	?	?	?	
Greater rhombicosidodecahedron						?

Table 5. Which polyhedra are possible using various module angles.

5 Spokes (Triangulation)

The triangulation (adding spokes) of a pentagon or a hexagon makes it possible to assemble polygons using modules with small characteristic angles. For example, it is possible to assembly all polyhedra with triangle, square, pentagonal and hexagonal faces in this way from a 45-module.

This technique also can be used also for squares and for modules with larger angle to obtain interesting models when triangulation is not necessary to assemble a polygon. Figure 8 shows the spoke technique used for a truncated tetradecahedron (the egg): the polyhedra skeleton, triangulation of hexagons, schema of modules, and a picture of a model.

6 Domes

We can triangulate pentagons and hexagons with the same module because the distance from the center to a vertex is smaller (pentagon) or equal (hexagon) to the length of the side. This condition does not hold in polygons of larger degree. We must build more complicated structures (domes) to obtain such polygons.

Figure 8. Examples of polyhedra made using the spoke technique.

To obtain octagons and decagons, we start with a square or pentagon (possibly with spokes). Then we double the number of sides using squares and triangles; the result is a *small dome*. When we use only triangles to double the number of sides, the result is a more complicated *large dome* (see Figures 9 and 10). The same technique can be used to obtain a hexagon from a triangle.

A construction method similar to the small dome was used by M. Matsuzaki [5] to build a large polyhedron model (the multimodular sphere). Figures 11 and 12 illustrate the small and large dome technique.

 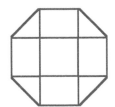 

Figure 9. Some graphs of small domes.

Figure 10. Some graphs of large domes.

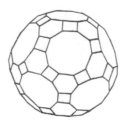
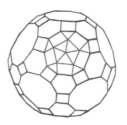

Figure 11. Illustrating the small dome technique.

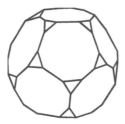
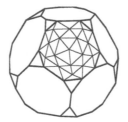

Figure 12. Illustrating the large dome technique.

7 Module Reversal

When large sinking angles prevent the assembling of a model, another technique can be used. Edge modules can be reversed along their "spine", making it a valley instead of a mountain. When reversed modules rise up instead of sink, the sinking angle is not important. Figure 13 shows examples of such a technique.

Figure 13. Examples of reversing an edge-module.

8 Our Goal Successfully Achieved (Almost)

We discussed the restrictions encountered when building polyhedra models with a class of edge modules. No single module can be used to build all regular and semiregular polyhedra. Three techniques were used to extend the polyhedra system of a particular edge module. These techniques can be joined and mixed to obtain new variations of polyhedra models. More examples of such models can be found at our web page:

http://www1.zetosa.com.pl/~burczyk/origami/index-en.html.

References

[1] Cuccia, L.A. and Lennox, R. B. and Ow, F.M.Y., "Molecular Modeling of Fullerenes with Modular Origami", *COET95 Second International Conference on Origami in Education and Therapy*, OrigamiUSA, New York, (1995).

[2] Fuse, T., *Unit Origami: Multidimensional Transformations*, Japan Publications, New York, (1990).

[3] Fuse, T., *Origami Spirals, Chikuma Shobo* (in Japanese), (1995).

[4] Gurkewitz, R. and Arnstein, B., *3-D Geometric Origami Modular Polyhedra*, Dover Publications Inc., New York, (1995).

[5] Kasahara, K. and Takahama, T., *Origami for the Connoisseur*, Japan Publications, New York, (1987).

[6] Macchi, P., "Icosaedro modulare bicolore A4", *Quadratto Magico 32*, (1993).

[7] Mitchell, D., *Mathematical Origami: Geometrical Shapes by Paper Folding*, Tarquin Publications, (1997).

[8] Simon, L. and Arnstein, B. and Gurkewitz, R., *Modular Origami Polyhedra*, Dover Publications, Mineolta NY, (1999).

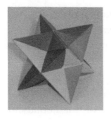

Using Graphs to Color Origami Polyhedra

Charlene Morrow

1 Introduction

I love modular origami, and I think this is mainly due to the opportunities it offers for posing coloring questions. These questions can then be used in many creative educational contexts. The explorations in this paper arise from my interest in "nicely" colored complex origami polyhedra. We will get to a more mathematical discussion of the meaning of "nice" in this context, but for now, understand it to mean symmetric (loosely speaking), balanced, or even just purposeful in some predetermined way. I had long been making fairly simple origami objects that are often described as polyhedra—tetrahedra, cubes, octahedra, dodecahedra, prisms (mostly boxes), and various stellated versions of some of these objects. More recently I began branching out into more complex polyhedra objects and was greatly motived in this direction upon seeing Thomas Hull's buckyball constructions. (For more information about buckyballs and how to fold and construct them visit Thomas Hull's website [10].

At the 1998 Origami USA Convention I attended a talk given by Hull, where he discussed using graph theory concepts to help color modular origami buckyballs. In the Fall of 1999 I visited Hull in his office at Merrimack College to learn more about the connections between origami and graph theory/polyhedra. After that I continued to study these connections and am reporting here what I've learned.

In order to talk about how to use graphs to color origami polyhedra, we will need to describe graphs, polyhedra, and what we mean by coloring. I am writing with an origami audience in mind, though I do hope that someone not very familiar with origami objects will be able to follow along and delve further into these topics with the references provided. My intent is to give some sense of the mathematical ideas that underlie the coloring problems I will be presenting rather than to give rigorous definitions of mathematical terms. I will begin by discussing the connection between polyhedra and certain origami objects.

Modular origami can serve as a great motivation for student research projects and investigations. The coloring questions posed here, and their solutions, use math that is easily understandable by students with little background in polyhedra or graph theory. This can give middle or high school students a chance to explore mathematical inquiry without the intimidation of complicated algebra or calculus. The quesitons we pose here can thus be used by educators to help lead students down these fun roads of exploration.

2   Thinking of Origami Objects as Polyhedra

Perhaps one of our first issues is to justify referring to some origami objects as polyhedra. The term polyhedra is very old; it has been defined and used extensively at least since ancient Greek times. Mathematically speaking, according to Peter Cromwell's recent book *Polyhedra* [3], there is no single agreed-upon definition of the term polyhedron; how you define it may depend on the historical era in which you find yourself. Greeks—Plato, at least—thought of these objects as solid and bounded by polygonal faces. Later generations thought of polyhedra as hollow, yet still bounded by polygonal faces, thus still giving a solid surrounding. Now some mathematicians accept the idea of polyhedra as frameworks, thus giving hollow objects with open faces, consisting only of vertices and edges. Regardless of your definition, though, there is still much that will hold true from one set of definitions to another: Cubes, whether solid, hollow, or open-faced, will have six "faces", eight vertices, and twelve edges. We also have to ask if polyhedra must all be convex (ball-shaped, more or less) or whether we can include concave (spiky or stellated) objects.

With regard to origami objects that we wish to call polyhedra, as in Figure 1, the operative idea is not to be too literal. Instead, think of modeling origami polyhedra with mathematical polyhedra. Look at your origami polyhedron and see what characteristics it shares with an abstract polyhedron. Use the idealized version to study and experiment with your object. Some origami polyhedra fit well with their idealized counterparts, but others have extra appendages

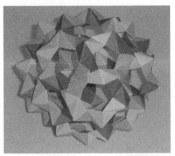

Truncated icosahedron (soccer ball):
Unit designed by Thomas Hull.

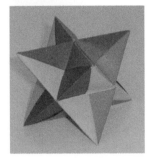

Six color dodecahedron: Three-color intersecting tetrahedra:
Unit designed by Robert Neale.   Unit designed by Tomoko Fuse.

Figure 1. Three examples of edge-colored origami polyhedra.

such as caps, indentations, and/or two-colored edges. Many capped (or spiky) origami polyhedra have an underlying structure that matches well with some convex polyhedron. And, of course, edges of origami objects, even if open-framed, are not one-dimensional as is the case with mathematical objects. My point here is not to be exhuastive in talking about the goodness-of-fit between mathematical and origami polyhedra, but to raise awareness about some of the aspects associated with thinking of origami objects in this way.

3   Platonic Solids

Before proceeding further, we will define the best known category of polyhedra, the Platonic solids, shown in Figure 2, which serve well as models for many origami polyhedra. Polyhedra are usually described and/or defined in terms

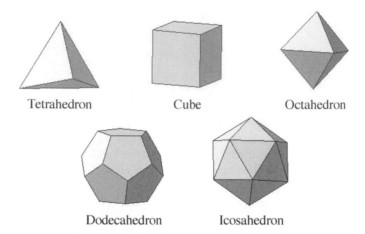

Tetrahedron   Cube   Octahedron

Dodecahedron   Icosahedron

Figure 2. The Platonic solids.

of their faces, vertices, and/or edges. The Platonic solids are characterized by each having equilateral/equiangular faces and equal vertex degree (each vertex has the same number of edges coming into it). Keep in mind that these objects are three-dimensional. Table 1 lists some important characteristics of the Platonic solids.

We can learn a great deal just by studying these characteristics. We can answer questions about the minimum number of colors it will take to color each edge so that no edges of the same color meet at a vertex. We can immediately see how many pieces of paper we will need to construct an object if each edge is constructed with one piece of paper. We can figure out the minimum number of colors that will be needed to make the edges of each face all different colors or to make no two adjacent edges the same color.

Name	Shape of each face	Faces	Vertices	Vertex degree (# edges at each vertex)	Edges
Tetrahedron	Triangle	4	4	3	6
Cube	Square	6	8	3	12
Octahedron	Triangle	8	6	4	12
Dodecahedron	Pentagon	12	20	3	30
Icosahedron	Triangle	20	12	5	30

Table 1. Characteristics of Platonic solids.

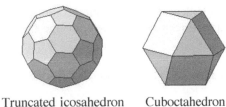

Truncated icosahedron   Cuboctahedron

Figure 3. Two examples of Archimedian solids.

4 Archimedian Solids

There is another class of polyhedra that serve as good models for many origami polyhedra: the Archimedian solids, two examples of which are shown in Figure 3. Important characteristics of these solids are listed in Table 2. The soccer ball is one of the more familiar items in this category. Cut off the vertices of an icosahedron and you get a soccer ball. When you cut off the vertices, you will virtually always get a figure that has pentagonal and hexagonal faces, but you must cut in a very precise way to get a regular soccer ball; that is one where all of the pentagons and hexagons are regular and all of the vertices are exactly the same, i.e., exactly two hexagons and one pentagon meet at each vertex.

5 Modeling Edge Colorings

We have already referred to the color of the edges of a polyhedron and we will want to talk more about this topic. As stated earlier, every origami polyhedron has an underlying mathematical polyhedron which represents its structure. If we look at what the units of our origami polyhedron correspond to on the mathematical polyhedron, we usually see that they correspond to the edges.

Name	Shape of each Face	Faces	Vertices	Vertex degree (# edges at each vertex)	Edges
Cuboctahedron	Triangle	8	12	4	24
	Square	6			
Truncated icosahedron (Soccer ball)	Pentagon	12	60	3	90
	Hexagon	20			

Table 2. Characteristics of two Archimedian solids.

Therefore, if we want to develop an interesting coloring of the origami version, we can look for interesting ways to edge-color the underlying mathematical polyhedron.

Unless otherwise stated, the reader should assume that the origami polyhedra being constructed are open-frame, where each edge is built from one piece of paper.

5.1 Coloring Challenges

Basic edge-coloring questions for origami polyhedra can be stated fairly simply. For example:

1. What is the fewest number of colors needed to make an origami cube if you want no edges of the same color meeting at a vertex?

2. What about the fewest number of colors needed to make a tetrahedron having no same color edges meeting at a vertex?

3. Can you make a dodecahedron with five colors such that each of the pentagonal faces is edge-colored with each of the five colors? If so, can you do this without having edges of the same color meet at a vertex?

4. How many different ways can you arrange three colors on a cube?

Without too much difficulty, readers can probably list many more coloring questions, most probably arising from their own origami constructions.

5.2 Polyhedral Graphs

Graph theory has many interesting and important applications to such things as communication networks, work assignments, scheduling, and coloring problems. Simply put, a graph is a collection of edges and vertices. The edges and vertices can be connected to each other or not and the edges can cross each other or not. There can even be more than one edge between two vertices. Figure 4 shows the graphs of the five Platonic solids, and Figure 5 shows the graphs of the two Archimedian solids pictured earlier. In graph theory there are coloring problems that concern the vertices, edges, or faces of a graph. It is edge-coloring that we will focus on in this paper. In particular we will look at some ways that graphs can be used to explore coloring problems with three-dimensional origami polyhedra. A simple example will be presented: Make the graph of a cube and explore various coloring schemes, using two, three, or four colors. Then more complex examples will be explored: 1) make the graph of a dodecahedron and use it to edge-color a thirty piece/three color origami dodecahedron such that each of the three different colors meet at each vertex,

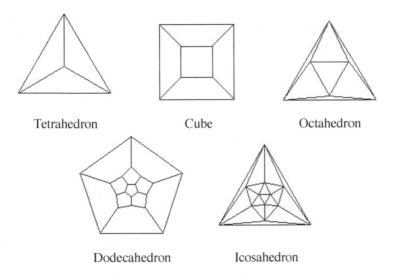

Tetrahedron   Cube   Octahedron

Dodecahedron   Icosahedron

Figure 4. Graphs of the Platonic solids.

or 2) make the graph of a soccer ball and color its 90 edges with the same stipulations as the dodecahedron mentioned above. Using a graph becomes particularly useful when applied to larger and more complicated origami polyhedra for which intuition about coloring is almost impossible.

In order to draw a graph of a polyhedron, we must represent each vertex and each edge in a two-dimensional format, for instance, on a piece of paper. The graphs of these objects can be drawn in ways other than those pictured, but it is essential that edge and vertex connections of the object are preserved in the graph. The physical shape of the faces is not preserved. In fact, if

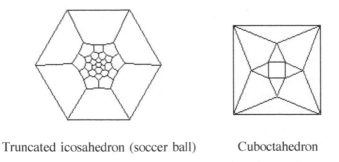

Truncated icosahedron (soccer ball)   Cuboctahedron

Figure 5. Graphs of two Archimedian solids.

you count the number of faces you can see, you might think that there is one missing—this is the face that is defined by the edges at the outer perimeter of the above graphs. A useful way of thinking about the connection between the polyhedron and its graph is to imagine taking the edges of one face and stretching the object out until it can lie flat on a piece of paper. Thinking of the polyhedron in the open frame sense might make this process easier.

One might notice that the graphs of Platonic solids can be drawn with no edge crossings. Graphs that have no edges crossing are called planar and have known mathematical characteristics that can be studied.

Polyhedral graphs give a way to focus on edge or vertex colorings in two-dimensional space, which is much easier than operating in three-dimensional space with complex objects. It quickly becomes impossible to keep track of colors on a three-dimensional object. Once we have drawn a graph for a polyhedron, we can simply color the edges or vertices with a plan in mind and see how it plays out. If we don't like the results or they do not work out as planned, we can start over.

I will offer one practical suggestion here. Though it is tempting to simply color the edges with colored pencils as you go, it is quite cumbersome to erase an entire edge if you want to change your mind (in fact, colored pencils do not always erase well). Instead I have found it much easier to simply choose a letter label for a color (R for red, B for blue, G for green, etc) and label each edge. When a coloring is found that works and/or is pleasing, the edges can be colored to get a clearer idea of the effect created.

Now, going back to some of the coloring questions raised earlier in this paper, let's ask how we can explore using other than random or trial-and-error approaches.

6   Hamiltonian Circuits and Coloring

One very interesting strategy that will work for many polyhedra is to find a Hamiltonian circuit, which is a particular kind of path through the vertices and edges. If this path can be found, it can provide a reliable coloring format. A Hamiltonian circuit is a path that starts at any vertex, travels along successive edges, and returns to the original vertex. The path must travel to each vertex once and only once. While all vertices are included on the path, not all edges are necessarily included. Figure 6 shows two examples of a Hamiltonian circuit (indicated by the thick line segments) on the graph of a cube.

Looking at the circuit we have identified on the cube, we see that it travels over eight of the twelve edges of the cube. Suppose we want to use three colors, R, B, and G, for our cube and we want each vertex to have each of the

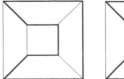

Figure 6. Two examples of a Hamiltonian circuit.

three colors. If we go along the circuit and color consecutive edges R and B in alternating fashion, we will have colored four edges R and four edges B. We also notice that each vertex has an R and a B edge coming into it (or going out of it). Since each vertex has only degree three, we now know that we can color the remaining edges—those that do not lie on the circuit—with the remaining color, G and each vertex is guaranteed to contain all three colors. Also the colors are distributed evenly over the twelve edges—four of each color. We can reliably three-color any polyhedron with vertices of degree three if we can find a Hamiltonian circuit. While an origami cube that is three-colored in this way is not hard to construct by observation or trial-and-error, the dodecahedron is much harder, making the use of a graph to color it much more appealing. Moving into more complex polyhedra like the truncated icosahedron (soccer ball) and beyond makes the use of a graph for planning a coloring scheme extremely effective.

Figures 7 and 8 show graphs of the dodecahedron and soccer ball each with a Hamiltonian circuit indicated. These graphs can be used to three-color origami polyhedra where one piece of paper is used for each edge in the manner described above, that is, alternately color the edges on the circuit with two of the colors; color the edges not on the circuit with the remaining color.

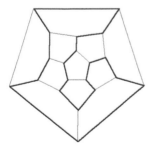

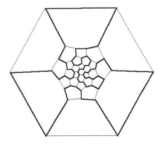

Figure 7. Hamiltonian circuit on the dodecahedron.

Figure 8. Hamiltonian circuit on the truncated icosahedron (soccer ball).

Another problem is to color the dodecahedron with a different number of colors, say six, that divides the number of edges (30) evenly. We could begin by coloring the circuit indicated in Figure 7 in four successive colors (this would mean five of each of the four colors). We still have ten edges not on the circuit, so we can use five edges in each of the two remaining colors. But can we distribute the colors over these edges so that a "nice" balance is maintained over the object? What might we mean by "nice"? This exploration is left to the reader.

Look at the Hamiltonian circuit identified on the graph of the octahedron in Figure 9. We see that because all of the vertices have degree four, there are two uncolored edges coming into each vertex. If we want vertices to have all different colored edges coming into them, we will need four colors. We can color the path alternately with two colors as for the cube, and then we would want to color the remaining two edges coming into the vertex with the two remaining colors. In looking at the situation more closely, we notice that there is no way to distribute the last two colors over the edges not on the circuit so that the edges coming into each vertex are all different colors. The circuit does not seem to work well in this situation. If we wanted only two colors at each vertex, we could color the circuit all one color and the edges not on the circuit another color.

Now look at Figure 10 to see what happens when we try a three-coloring of the octahedron, where we relax the condition that no edge color repeat at a given vertex. Color the circuit with all three colors: R, B, G, R, B, G. as illustrated. Then for each face that already has two edges colored, fill in the "missing" third color. You will have an octahedron that has the edges of each face having all three colors and the edges at each vertex being alternately colored in two colors. In addition, all of the combinations of three colors taken

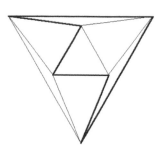

Figure 9. Hamiltonian circuit on an octahedron.

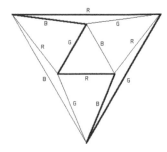

Figure 10. Coloring of an octahedron using a Hamiltonian circuit.

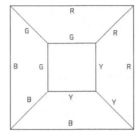

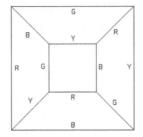

Figure 11. A three-coloring of a cube.   **Figure 12.** A different three-coloring of a cube.

two at a time are represented at the vertices with each of the three possible combinations occurring twice.

Another exploration for readers is to find a Hamiltonian circuit on the graph of the cuboctahedron and then see what results are produced by using an approach similar to that used in coloring the octahedron.

Finding Hamiltonian circuits is not the only way to make use of graphs as a coloring tool. Graphs can be used as a way to explore a wide variety of questions about coloring origami polyhedra without having to construct the origami objects themselves. Suppose I want to see how many different ways I can edge-color a cube using four colors. In this case, a circuit will not be an effective tool for producing the results I want, but I can still explore colorings on the graph of a cube as shown below: I can use trial and error, or perhaps I can use some knowledge I have about how many ways I can combine four colors if I take them in groups of three at a time. From my explorations, I can produce the two coloring layouts shown in Figures 11 and 12.

For the cube in Figure 11, there are four vertices where all three edges are the same color, one vertex for each color, and four vertices with edges of three different colors, each representing a distinct combination of four colors taken three at a time. For the cube in Figure 12, each face has edges of all four colors, and the edge color combinations at the vertices are again all combinations of four colors taken three at a time, but now there are two copies of each combination. Furthermore, the two copies of each combination occur at the two ends of each diagonal of the cube and are in opposite rotation to each other. Further explorations are encouraged. For instance, go beyond the Hamiltonian circuit that did not work for a nice four-coloring of the octahedron (discussed above for Figure 9) and find a layout that does work.

It can be tedious to fold a large polyhedron, so it becomes quite appealing to explore coloring schemes on a graph before choosing which ones to actually construct. Any polyhedron can be modeled with a graph. We have only dis-

cussed planar graphs (those with no edge crossings) in this paper, but objects
that do not have planar graphs could be modeled as well.

7   Transferring the Color Scheme to the Object

There is still the challenge of using the colored graph to guide the construction
of the origami object. The most important task is to keep track of the connection
between the piece of paper you are about to add to the origami construction and
where that edge is located on the graph. I begin by marking on the graph the
vertex where I will start my construction. I usually begin approximately in the
middle and work my way out in successive "rings." I construct this first vertex
out of paper. This means, for instance, that if I am making a dodecahedron
(open frame), I will have three pieces of paper interconnected at a vertex with
"edges" extending out in three directions. I mark this vertex with a paperclip so
that I will know which vertex corresponds to the marked vertex on the graph.
I then move to an adjacent vertex and construct it. For my dodecahedron, I
focus on getting the five edges of one pentagonal face made, which means I
will have five edges and five vertices, each with an edge extending out along
an adjacent face. In the beginning, it might be most efficient to mark off edges
on the graph when they are added to the origami piece. With experience, this
might not be necessary.

8   Symmetry and the Meaning of "Different"

I have used the words "nice" and "balanced" several times, but what do they
mean? When I use these words, I mean planning a color layout that is aesthetic.
What is pleasing can vary from person to person, but the idea of balance is
very important in both art and mathematics. This question could be explored in
a mathematical way by delving into the topic of symmetry. We will not cover
this topic here, but if these questions seem interesting, the reader is encouraged
to refer to [3] and [4].

9   Tools

I will leave readers with a final word about effective tools for exploring coloring
questions on graphs and with origami. Paper and pencil are an obvious choice,
but there are some good technology tools that are efficient and fun. The *Poly*
program [15] is a wonderful electronic catalog of hundreds of polyhedra. They
are categorized by type (Platonic solids, Archimedian solids, prisms, etc.) and

for each selection within a type, you can get three different three-dimensional views (which can be rotated), the net, and the graph. Each view can be copied and printed, giving ready access to a wide variety of graphs to use for coloring.

The *Geometer's Sketchpad* program [11] is an extremely useful tool for coloring graphs. Once a graph is constructed, the edges can be readily colored and changed at will, giving great flexibility for exploring colorings. An easy approach for making graphs is to copy and paste into *Sketchpad* a graph from the *Poly* program and then create vertices and edges right over the top of it, as if making a tracing. When the vertices and edges are complete, the underlying copy from *Poly* can be deleted and just the *Sketchpad* graph will remain. This is not a high level use of the geometric capabilities of *Sketchpad*, but it is a quick way to make a colorable graph.

In addition to elementary readings on graph theory and symmetry, the reference list offers sources for finding models and directions for origami polyhedra that lend themselves well to edge-coloring explorations. If the reader has had little experience with modular origami polyhedra, Fuse's *Unit Origami* [7] is a highly recommended starting point and an extensive source of models.

References

[1] Biggs, Norman L, Lloyd, E. Keith, and Wilson, Robin J., *Graph Theory: 1736–1936*, Oxford: Clarendon Press, 1976.

[2] Burczyk, Krystyna, *Polyhedra Pages*:
     http://www1.zetosa.com.pl/~burczyk/origami/galery1-en.htm.

[3] Cromwell, Peter, *Polyhedra*, Cambridge: Cambridge University Press, 1997.

[4] Farmer, David W., *Groups and Symmetry: A Guide to Discovering Mathematics*, Providence, RI: American Mathematical Society, 1996.

[5] Fiorini, S., and Wilson, R.J., *Edge-Colorings of Graphs*, San Francisco: Pitman Publishing Ltd, 1977.

[6] Franco, Betsy, *Unfolding Mathematics with Unit Origami*, Berkeley: Key Curriculum Press, 1999.

[7] Fuse, Tomoko, *Unit Origami*, Tokyo: Japan Publications, 1990.

[8] Gurkewitz, Rona and Arnstein, Bennett, *3-D Geometric Origami: Modular Polyhedra*, New York: Dover, 1995.

[9] Hart, George, *Virtual Polyhedra Encyclopedia and References*:
     http://www.georgehart.com/virtual-polyhedra/vp.html.

[10] Hull, Thomas, *Origami Mathematics Web Pages*:
     http://web.merrimack.edu/~thull/OrigamiMath.html.

[11] Jackiw, Nick, *Geometer's Sketchpad Software*, (From Key Curriculum Press): http://www.keypress.com.

[12] Kasahara, Kunihiko, *Origami Omnibus: Paper Folding for Everybody*, Tokyo: Japan Publications, 1988.

[13] Ow, Francis, *Origami Modulars*: http://web.singnet.com.sg/~owrigami/.

[14] Plank, Jim *Modular Origami Page*: http://www.cs.utk.edu:80/~plank/plank/pics/origami/origami.html.

[15] *Poly* computer program from Pedagoguery Software: http://www.peda.com/.

[16] Sanchez, Rosa, *Modular Origami Pages*: http://www.geocities.com/ rrrrrrosa/.

[17] Simon, Lewis and Arnstein, Bennett, *Modular Origami Polyhedra*, Los Angeles, CA: Bennett Arnstein, 1989.

[18] Trudeau, Richard J., *Introduction to Graph Theory*, New York: Dover Publications, 1993.

Circular Origami: A Survey of Recent Results

Eva Knoll

1 Introduction

For many years now, I have been studying systems of constraints in different design media. These studies, in turn, fuel my own creativity and inspire
me to produce artwork and theoretical papers in both art and mathematics.
My inspiration comes from many different media, including Islamic tiling,
Celtic interlace, Japanese paper-folding traditions, modern art, particularly the
Constructivist and Concrete movements, even natural phenomena and effects
of perception.

   In this particular instance, I will report on my experience so far using
circular paper for origami. This paper was deliberately separated into three
sections, starting with a short chronology of the discoveries I made in circular
origami. The middle part of the paper describes the mathematics that I derived
from these explorations. Finally, the last part reports on the results achieved in
mathematics education through the use of my discoveries and products derived
from them.

2 Circular Origami

The story of my explorations in circular origami begins with the coincidence
of two encounters I had in the summer of 1996. I met Chris Palmer, who
through his work reintroduced me to origami and demonstrated its exploratory

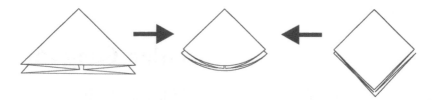

Figure 1. Same result from bird- and water-bomb base.

aspect. Shortly thereafter, I came across paper of appropriate quality for folding that was cut into circles. What a novel idea, and what an interesting set of parameters!

At first, of course, I used my circular paper like a square, folding it into 90 and 45 degree angles. This proved interesting only to a point. For one thing, the bird-base and water-bomb base amount to the same thing in circular paper (see Figure 1).

But the circle has some unique properties. Besides having infinite axes of symmetry, it lends itself well to certain applications of trigonometry. This might seem obvious in light of the fact that trigonometry is based on the geometry of the circle, but the results in this case are stimulating. In Figure 2, a fold is made that introduces an unexpected angle of 30 degrees without having to resort to measurement, approximation, or drawing. This is, of course, due to the fact that $\sin 30° = 1/2$!

This discovery led to some interesting results, particularly using a six-fold symmetric purse fold.

Further development was made using a tucking movement. This allowed me to create a concave or convex vertex in a specific point of the grid obtained from the earlier folds (Figure 3).

What happens if you apply this tucking system repeatedly to your paper?

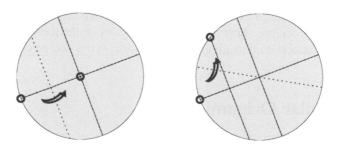

Figure 2. $\sin 30° = 1/2$.

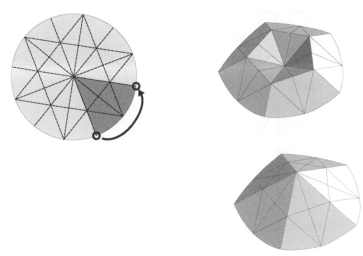

Figure 3. Concave or convex vertex.

Can you close up the shape? The answer here is yes, in many different ways. In the above drawing, 60 degrees were removed leaving five equilateral triangles exposed. It is also possible to remove 120 or even 180 degrees, or to leave a "vertex" of the grid unchanged. Figure 4 shows an example of a shape combining 6-vertices and 4-vertices. It can be made from a circle subdivided into 8 sections along the diameter.

This folding method can, of course, yield all kinds of results, but in certain cases, a new problem emerged. Because you can only generate subdivisions of the diameter into powers of 2 (2, 4, 8, 16, etc.), if you need, for example, 10 or 12 rows, you will use a large circle and much of its surface will have to be tucked away inside the shape, mostly at the closing point. In the case of the regular 5-6-5 deltahedron (Figure 5), you need 12 rows, leaving 2 extra

Figure 4. A 6-6-4-6-6 deltahedron.

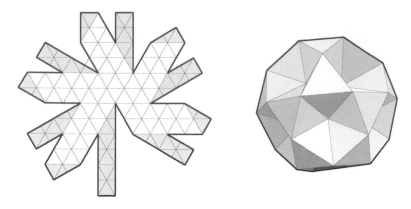

Figure 5. Geraldine (endo-pentakis-icosi-dodecahedron).

rows at each end (for a total of 16). The paper that needs to be tucked away becomes so unwieldy that I decided to remove some of it through cutting. This is a departure from traditional origami techniques, but I felt it justified by the interest of the result. The cutting technique itself is interesting because it mirrors the structure of the final shape (see also [2]). It consists of the removal of a wedge-shaped piece corresponding to half the angle that would be tucked away. This allows for the retention of a tab that will be used for assembly. In the case of Figure 5, the wedge has an angle of 30 degrees and the grayed-out areas correspond to the tabs.

The shape thus developed was so interesting in its own right that it became the centerpiece of my experimentations. Lacking the knowledge to determine its scientific name, I called it Geraldine (meanwhile John H. Conway was nice enough to identify it as an endo-pentakis-icosi-dodecahedron), and continued to experiment with it. The first step of this further study consisted in building a larger model using the same technique. Geraldine II was made out of a plastic sheeting used in buildings as a vapor barrier, shish-kebab sticks for rigidity, and hot glue. The finished shape was 1 meter (about 3 1/2 feet) in diameter (Figure 6). The initial circle was drawn using a pencil and a rope, and three people assembled the whole shape in a private home during a single afternoon.

Building the larger shape was an enriching experience, very different from the assembly of the small handheld model. The material reacted differently, yielding much more to gravity, and we had to work together using our whole bodies in collaboration, making the whole event much more of an all-encompassing experience, convincing us that we needed to make Geraldine even bigger!

Figure 6. Building Geraldine II.

Geraldine III (Figure 7) was made using a modular system based on kite technology, and the barn-raising event took place at Lanier Middle School (Houston, TX) in April 1999, and again at the Bridges: Mathematical connections in Art, Music and Science Conference (Winfield, KS) in July 1999 [1]. The modular system, a further step away from origami, had the added advantage of allowing us to build more shapes based on the same principle, including the shape of Figure 4, flexagons, twisted tubes, etc. (see [3]).

Circular origami has so far shown promising results. Even though with Geraldine, I have made a significant departure, finding a complex enough field

Figure 7. Building Geraldine III.

in its own right, I have continued to study circles and their folds. After all, its potential is probably as varied as its symmetries! In the next section, I will discuss more of the mathematical aspects of my results.

3   Some of the Mathematical Results

It would be difficult to exagerate the importance of the results obtained after the realization that $\sin 30° = 1/2$ opened the door to 6-fold symmetries in the folding of the circle. There are many more mathematical theories and theorems that can be illustrated through my developments in circular origami.

In Figure 3, we see the beginnings of an illustration of the Euler characteristic (faces - edges + vertices = 2) and the angle deficit theorem ([4]) according to which any closed polyhedron without holes in it (no donuts allowed) will have a total missing angle of two whole turns (i.e., 720°). Using the example of Figure 4, one can see that there are only six vertices where material was tucked away (the six 4-vertices). Furthermore, at each vertex, two whole equilateral triangles (so 2×60 degrees) were removed, for a total of $6 \times 2(60) = 720$ degrees, two whole turns! In the case of Geraldine, there are twelve 5-vertices (in the middle of the dimples) and the rest are 6-vertices, so $12 \times 1(60) = 720$, again, two whole turns. But what happens when some of the vertices have more than six triangles meeting? Taking the example of the stella-octangula (which can also be built from a circle, albeit a little differently; Figure 8). The shape comprises eight 3-vertices and six 8-vertices. The calculation would therefore be: $8 \times 3(60)6 \times 2(60) = 720$. In this case, the second term in the calculation is negative because it is an angle increase when we are calculating the deficit.

Returning to Geraldine, it was soon apparent that she had the same symmetry and general structure as the regular icosahedron (in fact, if you bump out

Figure 8. Stella-octangula.

all the dimples, you obtain an icosahedron of "edge-length" equal to 2). It is also possible to build a "dimpled" dodecahedron using the same system. This observation prompted me to ask the following questions: What is the generating principle of all these related shapes, and are there more similar ones? The second part of the question can be answered in the positive simply by observing that the "soccer ball" can be similarly dimpled on its pentagonal faces to obtain, again, a polyhedron of the same family (made of only equilateral triangles and possessing the icosahedral symmetries). Trying to answer the first question took us, via vector geometry, straight to number theory ([5]). Using these varied mathematical tools, we were able to demonstrate that there is, indeed, a generating principle, and also that the shapes generated have some qualities that are constant. An example of this is that in the case of the icosahedral family, the shapes will always be composed of a multiple of 20 faces. Further, this multiple corresponds to the square of the shortest surface distance (always a whole number) between two "adjacent" 5-vertices.

There are more observations of a mathematical nature to be made here, including such challenges as color distribution, whether the previous is applicable to other symmetry groups, etc. The most important aspect to note here are the parallels that can be made between folding and understanding mathematics. Having understood this, it was easy to envision a connection with mathematics education. Having worked with a doctoral student in mathematics at Rice University, I was able to transfer these experiments into a school setting, witnessing first-hand the pedagogical worth of the exercise.

4 Education

The education applications of the preceeding experimentation began in 1999 through the Rice University School Math Project, a resource for mathematics teachers in the community in Houston, TX.

Early sessions took place at Lanier Middle School. The first session consisted of the building of an icosahedron from a circle using the paper-folding method referred to in the first part of the paper. After a few exercises aimed at practicing using the triangles, the second part of the event consisted of building Geraldine in one of the courtyards of the school. The two parts together were significant because they gave the students the opportunity not only to handle the shapes themselves, but also to build them in such a way that the process reflected the final object at every step. This is a definite departure from a typical exercise where the teacher gives the students a model of a polyhedral net that mysteriously assembles into the shape the teacher promised. It is not possible to overemphasize the importance of this aspect. If the students can follow the reasoning process from beginning to end, they never need to take the teacher's word for it. Furthermore, this type of exercise, particularly the part with "life-sized" triangles goes a long way toward reconnecting the abstraction of mathematics with reality. Finally, the manipulatives used in these exercises are ideal for discovery learning as they let the students experiment, and more specifically, they can experience the thinking process with their whole body, rather than only their fingertips in the case of a handheld model or than their eyes only, in the case of a film or computer program.

The circular origami exercise and the large triangles have been used successfully at various levels from K-12 all the way to an undergraduate mathematics environment, both with advanced and at-risk students. Work is set to continue in this direction, with a particular emphasis on documenting the various lessons, as well as, developing further applications.

5   Conclusion

The story of my experience with circular origami is by no means ended at this point. The success of the applications both in mathematics and education demonstrates the potential for further development. All that can be said is: 'til next time!

References

[1]   Knoll, Eva, Morgan, Simon, *Barn-Raising an Endo-Pentakis-Icosi-Dodecahedron*, Bridges Conference Proceedings, Winfield, Kansas, 1999.

[2]   Knoll, Eva, *From circle to icosahedron*, Bridges Conference Proceedings, Winfield, Kansas, 2000.

[3]   See the some more pictures at the ISAMA Website: http://www.isama.org/.

[4]   Morgan, Simon, Knoll, Eva, *Polyhedra, learning by building: design and use of a math-ed tool*, Bridges Conference Proceedings, Winfield, Kansas, 2000.

[5]   Knoll, Eva, *Decomposing Deltahedra*, ISAMA Conference Proceedings, Albany, NY, 2000.

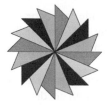

In Praise of the Papercup: Mathematics and Origami at the University

Emma Frigerio

1 Workshop on Mathematics and Origami

Last year, I had the opportunity to teach a workshop on math and origami for students majoring in primary education. These students, in addition to basic courses in language, math, and science and so on, and courses in psychology and pedagogy, are required to take some workshops of a more practical and/or cross-disciplinary nature, to be chosen from a wide offering.

Of course, in grade school, origami can be used to enhance different kinds of skills and for many different purposes. But I am a mathematician and my workshop was in the math area, so I concentrated my attention on various mathematical aspects of origami. Now, it is true that when you do origami, you also do, whether you notice it or not, some kind of mathematics and you eventually absorb it. For instance, once you have folded a preliminary base from a square piece of paper several times, then you know that, when you halve the side of a square, the area is divided by four. So, if you received, as I did last year, a pizza ad saying:

| maxi | diameter 46 cm | 18000 lira |
| mezza | diameter 23 cm | 10000 lira, |

you would notice immediately that the name "mezza", which means half, is misleading and the price is unfair: If with a maxi pizza, you can feed four

291

people, then with a mezza, you would expect to feed two, while you actually can feed only one, and at a high cost! This example is quite obvious, but in general, the geometry underlying origami is more subtle; my main goal was to make it more evident.

Twenty-six students took part in the workshop, which was organized in four four-hour sessions, plus a final meeting, a few weeks later, to discuss and evaluate their projects. With only two exceptions, they had no previous knowledge of origami, so I had to start from the very beginning. Hence the pace had to be fairly quick, if I wanted to bring them to a point where they could continue by themselves. Also, had I been too systematic, I would not have covered enough material so as to convey the main ideas and to give some hints on other possible developments, but had I been too informal, I would not have given them a method of using origami to do math.

2   Papercup Model and Modules

The traditional papercup model proved to be a good starting point, for three reasons: (1) it is easy to teach and learn (hence useful to introduce basic folding instructions); (2) many mathematical observations on its crease pattern can be done; and (3) some easy modules can be derived from its folds.

First, I taught the students how to fold the papercup by imitation, then I introduced the standard folding notation and showed them Figure 1.

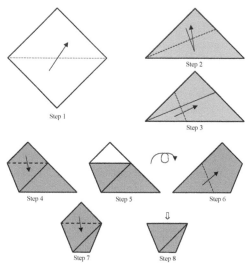

Figure 1. How to make the traditional papercup.

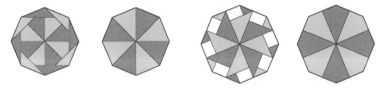

Figure 2. The fronts and backs of two coaster designs.

I then introduced modular origami and taught them two possible ways of assemblying eight modules obtained by folding the papercup up to Step 5. Both sides of the octagonal coasters which result when alternating two colors are depicted in Figure 2.

At this point you can ask the students to find other ways of assembling the same modules, or other modules derived from the papercup folds, or ways to assemble together both right- and left-handed modules. Figure 3 illustrates some possibilities. Proceeding clockwise from the top left, we have:

1. Same modules, different color pattern.

2. Modules based on Step 3, assembly similar to the second.

3. Modules based on Step 2.

4. (Bottom four.) Right and left modules together (one color for right and another for left modules or two of either type for either color; fronts and backs shown).

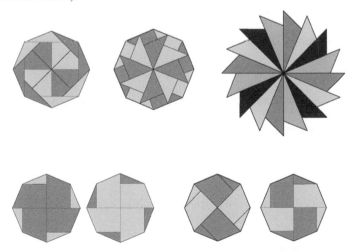

Figure 3. More examples of modular designs.

WORKSHEET 1

PAPERCUP

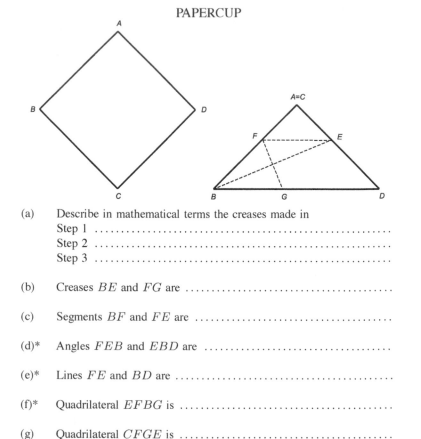

(a)   Describe in mathematical terms the creases made in
Step 1 ....
Step 2 ....
Step 3 ....

(b)   Creases BE and FG are ....

(c)   Segments BF and FE are ....

(d)*   Angles FEB and EBD are ....

(e)*   Lines FE and BD are ....

(f)*   Quadrilateral $EFBG$ is ....

(g)   Quadrilateral $CFGE$ is ....

(h)   Which relationship holds between angles CEF and FEG? ....

(i)   Crease made in Step 6 passes through point E because ....
....
....
....

     Mark this crease on the figure and let H be its intersection with BD.

(j)   Quadrilateral $EFGH$ is ....

*   Try to prove your assertion.

3 The Worksheet

In these notes, I want to show in detail how much geometry is hidden in the crease pattern of the papercup. Worksheet 1 is a translation of the worksheet I had prepared to guide the students through the process of discovery. The basic questions are of three types:

1. Identify the mathematical meaning of some creases.

2. Recognize some facts (equality of segments or angles, parallelism of lines and so on).

3. Prove some of them (the ones marked with a *).

In fact, although these people will teach only intuitive geometry, I believe that some training in theorem proving is important for a better understanding of the subject; on the other hand, I felt it was too much to ask them to prove everything. What follows are the answers and some comments:

(a) Most students described the crease made in Step 1 as "diagonal", which is, of course true, but also, in some sense, false (see the end of this paragraph). The crease made in Step 2 was correctly identified as "angle bisector" by all students. The description of the crease made in Step 3 is obviously more difficult. Most students answered "diagonal of rhombus $EFBG$"; what is wrong with this answer is that you do not have its vertices F, G before you fold (and you don't know it is a rhombus yet). The correct answer is "perpendicular bisector of BE", since any point of the crease is equidistant from B and E, but only few students gave it. This is the best answer to the first question as well: What is actually folded in Step 1 is the perpendicular bisector of AC, which happens to be the diagonal BD.

(b), (c) Most students answered correctly "perpendicular" and "congruent" to these two questions, which were meant to help them with the previous question, but they did not get the hint.

(d) The two angles are congruent.
   Proof: $\angle FEB = \angle FBE$ (they coincide with fold FG),
   $\angle FBE = \angle EBD$ (BE is angle bisector), hence
   $\angle FEB = \angle EBD$.

(e) The two lines are parallel.
   Proof: The alternate interior angles FEB and EBD are congruent by (d).

(f) The quadrilateral is a rhombus.

Proof: The triangles FBE and GEB are isosceles (FG is the perpendicular bisector of BE) and congruent (base in common and congruent angles), hence all sides of $EFBG$ are congruent. Different proofs are also possible.

Most students had claimed the quadrilateral is a rhombus in (a); nevertheless they tried to give a reason here, and some succeeded.

(g) The quadrilateral is a right trapezoid.

Proof (not requested, but not difficult): $\angle CEG = \angle CEF + \angle FEG = 45° + 45° = 90°$, hence lines FC and GE are parallel, being both perpendicular to line CE.

(h) The two angles are congruent, as shown above (this allows Step 4, i.e., the insertion of the flap CFE into the pocket made in Step 3).

(i) The crease is the perpendicular bisector of FD; moreover $FE = FB$, and $FB = ED$ (the trapezoid $EFBD$ is isosceles, since its base angles are congruent), hence $FE = ED$, therefore E is a point of the crease.

The answer is not so easy, but some people got it right.

(j) The quadrilateral is an isosceles trapezoid.

Proof (not requested, but easy): FG and EH are the short diagonals of congruent rhombi $EFBG$ and $EFHD$.

Of course, it is possible to ask other interesting questions, for instance: find the ratio between the areas of CFE and $EFGH$ (answer: 1/2).

4   Conclusion

Starting in this way, rather than from a more formal introduction to geometric construction via origami, I was able to cover a good amount of material and to keep students' interest alive.

Elsewhere in the workshop, I used a similar approach and I tried to alternate practice (folding) to theory (such as asking them questions like: Why does this crease pass through that point? Why are these three points aligned after folding? Can this model be folded from any rectangle, from only one rectangle or from many but not all? What is the ratio between the area of the model and the area of the sheet?).

I prepared two more worksheets, one on 60° and 30° angles from a square, inspired by Kasahara and Fuse, and another, based on a letterfold by Strobl, which explores the geometry of the so-called silver rectangle, i.e., whose sides are in the ratio of $1:\sqrt{2}$, the standard size for paper in Italy and in most European countries.

The other major topic of the workshop was regular polyhedra. The division of the material in the four sessions was roughly as follows:

1. Introduction
   Papercup
   Octagonal coasters
   Practice on basic folds

2. Square from a rectangle
   Rectangles and similarity; the silver rectangle
   Sonobe's 90° module; cube

3. Fuse's and Macchi's 60° modules (from square and from 1:3 sheets);
   tetrahedron, octahedron, icosahedron, some stellations
   Fujimoto's approximate division of a rectangle into 3, 5, 7 equal parts

4. An origami geometry theorem on triangles
   Simon's module for the dodecahedron
   Concluding remarks and teaching tips

Evaluation of the students was based on the three worksheets and on some projects. They could choose from a list including:

1. Make folding diagrams of a traditional model and prepare a worksheet for mathematical observations on it.

2. Find how to divide a rectangle into 11, 15, or 17 equal parts.

3. Construct some polyhedra fulfilling assigned color requirements.

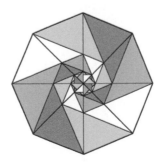

Figure 4. Judy Hall's coaster version.

All of the students worked with pleasure and learned some math; most of them showed some kind of skills (intuition, formal reasoning, drafting, folding ...) and did more than I had asked them to do.

In conclusion, I think this experience was useful and gratifying both for them and for me and I will do it again this year and, hopefully, in future years.

Afterward

At the 3OSME Meeting, after my presentation, Judy Hall found yet another way of assembling eight "standard" modules and got the beautiful coaster depicted in Figure 4.

Using Triangular Boxes from Rectangular Paper to Enrich Trigonometry and Calculus

V'Ann Cornelius and Arnold Tubis

1 Introduction

It has been recognized in the USA for almost one hundred years (e.g., Row [6]) that paper-folding may be used to discover and demonstrate relationships among lines and angles of Euclidean plane geometry. During the past 25 years, several monographs that incorporate origami and informal plane geometry, mostly focused on middle school mathematics, have been written (e.g., Olsen [5], Johnson [2], Serra [8, 7], Jones [3], Franco [1], Youngs and Lomeli [9]). These efforts have been designed to promote the use of origami as a means of informally introducing the main concepts, constructions, and implications of plane geometry as a preliminary to the traditional theorem-proof geometry course in the ninth or tenth grade.

In the works just cited, except for the books by Franco and by Youngs and Lomeli, little or no effort is made to incorporate both folding and quantitative mathematical analysis of typical origami models. It is our hope that the synergy involved in the introduction of origami as a method of making interesting objects as well as a vehicle for introducing and practically applying geometric, trigonometric, and calculus concepts can be of great pedagogical value. As an illustration of our approach, we describe in this paper the origami and the associated mathematics of folding a simple, but interesting class of isosceles triangular boxes with *any* vertex angle θ ($0°$ to $180°$) from rectangular paper of arbitrary length L and width W. It will probably be surprising to students that boxes with any vertex angle can be folded in principle. (Of course, it would be physically impossible to actually fold boxes for certain ranges of the parameters L, W, and θ.)

2 Folding Procedures

The first author had originally designed right-apex-angle $(45° - 45° - 90°)$ and equiangular/equilateral $(60° - 60° - 60°)$ pie containers from arbitrary rectangular sheets. It was then noted by the second author that the folding procedure could be generalized so that a box with any arbitrary apex angle θ could be folded using well-defined landmarks.

The general procedure is outlined in Figures 2 and 3. The angle ϕ in Step 2 is equal to $\theta/2$, where θ is the desired vertex angle. (It will be instructive for students to apply the binary folding algorithm of Robert Lang (see [4]) in approximating the desired value of ϕ.) Step 4 shows that choice of θ determines H, where H is the height of the folded box. The forming of the triangular face of the box starts in Step 8 by rotating edge a about point D until it meets line a' and then making valley crease c. It is easily seen that the constructions of the $(45° - 45° - 90°)$ and $(60° - 60° - 60°)$ boxes are special cases of the one just described, in which some of the folding steps in the general case (such as the forming of valley fold m in Step 5) are not necessary. Note also that it is not necessary for L to be greater than W.

For $\theta = 180°/5 = 36°$, the entire bottom face of the box is exactly five layers thick, and for smaller apex angles, extra pleating of the layers of the bottom face is required. This extra pleating is not shown on the instruction diagrams, and in any case, the box becomes progressively less sturdy and more difficult to fold (and hence less satisfactory as a practical container) for apex angles $\theta < 36°$.

3 Analysis of Height, Base, Area, and Volume

The relationship between the parameters L, W, H, and θ is easily determined by analyzing Step 4 of Figure 2, which is annotated in Figure 1.

$$L - (W/2)\cos(\theta/2) = 4H. \tag{1}$$

Using the fact that $0 < \theta < 180°$, i.e., $0 < \cos(\theta/2) < 1$, we see also that

$$L - (W/2) < 4H < L. \tag{2}$$

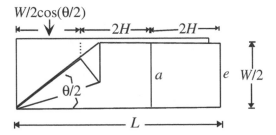

Figure 1. Diagram for determining the relationships between L, W, H, and θ (Equations (1) and (2)).

Since $H > 0$, then $L > W/2$. However, this condition is not a restriction on the proportions of the paper. If $L \leq W/2$, then $W \geq 2L > L/2$, in which case we interchange the labels "length" (L) and "width" (W). We assume in the remainder of the paper that $L > W/2$.

An interesting exercise for students would be to specify the height H of the box for given dimensions of the starting rectangle, and then determine the requisite θ by folding with well-defined landmarks. The folding steps are not shown, but may be easily determined from the steps shown in Figures 2 and 3. It should be noted (or students will soon discover empirically) that for a range of H values, the box-folding may be impossible. We note from Equation (2) that the specified H must be $< L/4$.

Deriving equations for the area A of the triangular face and volume V of the boxes is an interesting trigonometric exercise. First, the area of the triangular face: Noting from the folding process (see Step 10) that the altitude h is $L - 4H$ and that $\tan \frac{\theta}{2} = \frac{1/2 \text{ base}}{h}$, we see that $A = \tan \frac{\theta}{2}(L - 4H)^2$. Use of Equation (1) and the double-angle formula simplifies this to

$$A = (W^2/8) \sin \theta.$$

Surprisingly, this formula is independent of the length L. The volume of the box is then

$$V = (W^2/8) \sin \theta H = (W^2/32) \sin \theta (L - (W/2) \cos(\theta/2)).$$

This suggests another exercise: Given L and W, find the maximum values of A and V, and the values of θ for which A, V are maximized. Pre-calculus students may numerically evaluate A and V for various values of θ to determine the maxima. Calculus students can easily maximize A by using

$$\frac{dA}{d\theta} = 0 \Rightarrow \cos \theta = 0$$

and

$$\frac{d^2 A}{d\theta^2} < 0 \Rightarrow -\sin\theta < 0$$

so that $\theta = 90°$ and thus the *exact* expression for the maximum area of the box face is

$$A_{\max} = W^2/8.$$

To maximize V, we use

$$\frac{dV}{d\theta} = 0 \Rightarrow \cos\theta \left[1 - \left(\frac{W}{2L}\right)\cos(\theta/2)\right] + \left(\frac{W}{4L}\right)\sin\theta\sin(\theta/2) = 0.$$

For θ values between 0 and 180°, $\sin\theta, \sin\frac{\theta}{2}$, and $\cos\frac{\theta}{2}$ are positive. Because

$$L > W/2, \frac{W}{2L} < 1$$

and so

$$1 - \left(\frac{W}{2L}\right)\cos(\theta/2)$$

is positive as well. This implies that $\cos\theta < 0$, or $90° < \theta < 180°$. An exact value for θ is not attainable unless one uses a computer algebra system, *and is* given particular values of L, W. (For 8.5" × 11" paper, we find that $\theta \approx 101°$ when $L = 11$ and $\theta \approx 113°$ when $W = 11$.)

Advanced calculus students familiar with Taylor series can use the standard degree-two expansions of $\cos\theta, \sin\theta$ to achieve an approximate value (to within a few degrees) for θ. (This could be an ongoing investigation for a calculus class!) Once they obtain numerical values for θ given fixed L, W, students can verify that

$$\frac{d^2 V}{d\theta^2} < 0.$$

4   Conclusion

The mathematics involved in analyzing our triangle boxes is surprisingly rich. Investigating this simple class of origami containers is a novel and useful means of involving high school students with practical applications of trigonometry and calculus. While typical study questions for student explorations of these containers are not presented, these can easily be developed. It is hoped that this paper will stimulate teachers to work out similar projects using other origami models whose designs are both attractive and amenable to practical analysis with well-known tools of high school mathematics.

1. Fold rectangle in half.

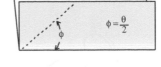

2. Make a valley fold at desired angle $\phi\ (= \theta/2)$ with respect to the creased edge formed in Step 1.

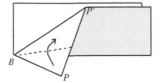

3. Make a valley fold by laying edge BP along BP' and creasing.

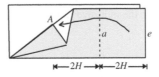

4. Bisect the perpendicular distance between point A and right edge e by making a valley fold a.

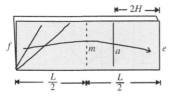

5. Undo the folds make in Steps 2–4. Place left edge f on right edge e and make a valley fold m.

6. Make a valley fold a' aling line a (hidden) with the top two layers.

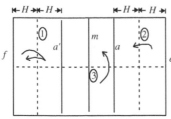

7. Open the paper. Make a valley crease 1. Unfold. Make valley folds 2 and 3 in succession.

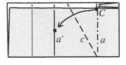

8. Make a mountain fold a and valley fold c using all layers by bringing point C to line a' while pivoting line DC counter-clockwise about point D.

Figure 2. Page 1 of instructions for the triangle box.

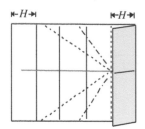

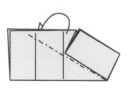

9. Make the indicated mountain fold.

10. Open up the model so that only valley fold 2 of Step 7 is made. Use existing creases to make the indicated mountain and valley folds (see Step 11).

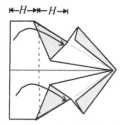

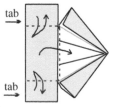

11. Make a valley fold to form the base wall of the box.

12 Make the two valley creases and valley fold. If $\theta < 60°$, the extra paper near the tabs should be tucked inside the wall of the box.

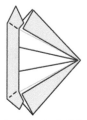

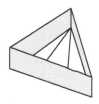

13. Tuck these "tabs" into the "pockets".

14. Complete.

Figure 3. Page 2 of instructions for the triangle box.

References

[1] Franco, Betsy, *Unfolding Mathematics with Unit Origami*, Key Curriculum Press, Berkeley, CA, 1999.

[2] Johnson, Donovan A., *Paper Folding Geometry*, J. Weston Walch, Portland, ME, 1976.

[3] Jones, Robert, *Paper Folding: A fun and effective method for learning math*, LWCD Inc., St Louis, MO, 1995.

[4] Lang, Robert J., "Folding Proportions", in *The Paper*, Origami USA, Issue 70 (August 2000), pp. 10–12.

[5] Olsen, Alton T., *Mathematics through Paper Folding*, National Council of Teachers of Mathematics, Reston, VA, 1975.

[6] Row, T. Sundara, *Geometric Exercises in Paper Folding*, edited and revised by Wooster Woodruff Beman and David Eugene Smith, Open Court Publishing Company, 1905; reprinted by Dover Publications, New York, 1966.

[7] Serra, Michael, *Discovering Geometry: An Inductive Approach, Second Edition*, Key Curriculum Press, Emeryville, CA, 1997.

[8] Serra, Michael, *Patty Paper Geometry*, Key Curriculum Press, Berkeley, CA, 1994.

[9] Youngs, Michelle and Lomeli, Tamsen, *Paper Square Geometry: The Mathematics of Origami*, AIMS Education Foundation, Fresno, CA, 2000.

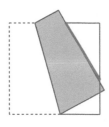

Fold Paper and
Enjoy Math: Origamics

Kazuo Haga

"ORIGAMI" has become an international word, derived from the Japanese word "origami". They differ somewhat in meaning as well as pronunciation. The accent of the former falls on the third syllable (=ga) while the latter on the second one (=ri), that is, ori[ga]mi and o[ri]gami; most Japanese pronounce it with a nasal sound.

In Japan, origami usually refers to a handicraft hobby mainly for children. Thus almost all books on origami are found in the juvenile section of the bookstore even though some of them are for enthusiasts.

I felt it was necessary to give a new name to describe the genre of scientific origami: I proposed the term *origamics* at the *2nd Origami Science Meeting* in 1994. The term origamics is composed of the stem, origami, and the suffix, ics, which is often used to stand for science or technology, as in mathematics.

In this paper, I will present three origamics topics from a folded square or rectangular sheet of paper.

1. The first step of folding induces an insect face pattern.

2. Haga's Theorem enables a standard rectangular paper to divide its length into an odd number of equal parts.

3. An arbitrarily made mother line bears 11 wonder babies.

1 The First Step of Folding Induces an Insect Face

The first origamics example shows how a single folding brings forth unexpected evolution. The reader should try this example while reading, experiment, and enjoy this world of mathematical wonders.

Please take out a piece of square paper which is colored (say, blue) on one side and white on the other. Place it on a table with the white side face up.

First, fold the paper once, however you like. You may fold only once arbitrarily! Random folding is unreproducible. Thus it isn't done in usual origami folding, nor is it done in mathematical folding. Some readers may be waiting for the next instruction. There is no need for further folding; the one fold is all.

By folding once, you have a part of the colored face which has come to the surface, from the back side to the front side. We will call this part TUP, which stands for "Turned Up Part."

Please pay attention to the TUP's shape. How many corners does your TUP have? Or rather, how many sides does yours have?

You probably have a triangle or a quadrilateral as shown in Figure 1. Might you get anything else?

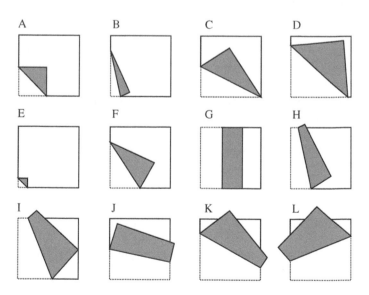

Figure 1. Examples of TUPs.

Let's study triangles and quadrangles first. Take a new sheet of paper of the same color, and make a different shape by only one folding. If your TUP was a triangle at first, try to fold the new paper to make a quadrangle. If you had a quadrangle at first, try to fold the new paper to make a triangle. Compare the triangle and quadrangle you made. You may fold more sheets to find conditions which make triangles or quadrangles.

What determines whether we obtain a triangle or quadrangle TUP?

A 15-year-old junior high school student in Japan described it as follows: When we unfold the square paper after folding once, there is a fold line (or a crease). When the TUP is a triangle, the line connects two adjacent sides of the square. When the TUP is a quadrangle, the fold line connects two opposite sides. While this is true, to be more accurate, we should add that, "When at least one end of the fold line falls on a corner of the square, it makes a triangle."

Another junior high school student phrased it as follows: When the folding moves only one of the four corners (vertices) of the square, it makes a triangle, and when the folding moves two corners, it makes a quadrangle. (Grown-ups could not have easily found such an idea, possibly because they take things too seriously.) This student's discovery seems obvious, but I think the student found the law of movement of corners through repeated folding and unfolding. This kind of experimentation process is important to science.

When this subject first came to my mind, the explanation I found was a modification of the second student's. That is, whenever we fold an origami, at least one corner moves. The new position of that corner determines the shape of the TUP. Let's pay attention to one corner and let it be "the main vertex (corner)." Please choose one corner as the main corner of your square paper and mark it. I usually mark the bottom-left one as the origin of coordinates. My assumption was, if the surface of the square paper has an invisible map of some divided towns, where the main corner has moved determines the shape of the TUP. This conception makes an interesting development.

Mark a new square with some dots at random as in Figure 2. Bring the main corner (marked with a small circle) onto each dot and look at the shape of TUP. Mark a dot which made a triangle with 3 and a dot which made a quadrangle with 4. Then, we can examine the distributions of each kind of dot. However, when we want to make a more exact map of the triangle town and the quadrangle town, we must examine infinitely many dots. To be more systematic, we should consider a grid of dots. (We could use a graph paper cut into origami-sized squares.) Then we can define each intersection of vertical and horizontal lines by a coordinate point. We must then classify all intersection points 3 (triangle) or 4 (quadrangle) (see Figure 3, left). In my class, I found that some students drew a boundary between the triangle and the quadangle towns like stair steps (as in Figure 3, part A). One should instruct

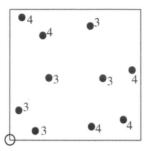

Figure 2. Random dots for TUPS.

students to draw a smooth line as the boundary, as this method has an accurate boundary between the triangle and quadrangle regions in the limit case (Figure 3, part C).

So, let's go back to the second student's idea. We move the main vertex and it makes a triangle when the other three vertices remain still. It makes a quadrangle when one vertex of the three also moves. We have to look for the boundary between the triangle and the quadrangle regions. When the main vertex traverses the face of paper, it is accompanied with vertical and horizontal sides folding over. When the whole length of a side has been moved, another vertex is about to be moved. Therefore, if we fix the next corner to the main vertex with a finger while sliding the main vertex, it draws an arc, a part of circle, with the radius of the side length of the square. We can trace the arc on the paper. If you fixed the vertically adjoining corner first, fix the horizontally adjoining corner next and draw an arc likewise.

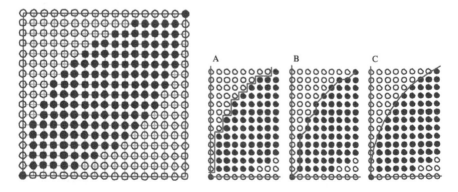

Figure 3. Grid samples indicate the triangle and quadrangle towns.

Figure 4. The triangle and quadrangle regions (the main vertex is the bottom-left corner).

As the result, you will find that those dots which make triangles are in a lens-like shape region along a diagonal of the square and those making quadrangles are outside of this region.

Figure 4 shows the arcs drawn with a pair of compasses. You can confirm that when we fold the paper bringing the marked lower-left vertex into the lens-shaped region enclosed by the arcs, the TUP makes a triangle. When we bring the marked corner outside of the lens area, the TUP makes a quadrangle. Let's call the inside and outside area the triangle region and the quadrangle region, respectively. The border line belongs to the triangle region. We should also study what will happen if each of the different corners of the square are considered to be the main vertex. The diagonally opposite vertex makes the identical figure as in Figure 4, while an adjoining vertex makes a reversed lens shape region. Therefore, when we bring any vertex of the four into the area where two lenses overlap, the TUP always becomes a triangle. As it were, it can be called the "absolute triangle area," as shown in Figure 5.

However, to fully understand this problem, Figure 4 is insufficient. Please look back to Figure 1, parts K and L. A quadrangle is produced, but the main vertex in the bottom-left corner has jutted out of the paper, that is out of the region in Figure 4. In other words, some quadrangle regions need more space

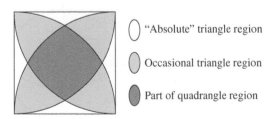

◯ "Absolute" triangle region

◖ Occasional triangle region

⬤ Part of quadrangle region

Figure 5. Triangle and quadrangle regions when any corner may be used at the main vertex.

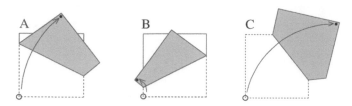

Figure 6. Examples where the main vertex is moved outside the square.

around the original square sheet (see Figure 6, A and B). In order to examine the whole quadrangle region, place a square sheet of paper on top of a larger piece of paper, and move the main vertex while holding still another part of the square to avoid its sliding. By experimenting, you will see that not only quadrangles, but also pentagon TUPs appear when the main vertex is moved outside the square (see Figure 6, part C).

Let's remember the original problem: Place a square sheet on the table, with colored face down, and fold the paper once. Then, examine the shape of the colored face that has come to the surface, which we call a TUP. It naturally includes the type of foldings shown in Figure 6. Thus the TUP can become a pentagon.

Now, we have found that when two vertices move, a quadrangle appears; and when three vertices move, a pentagon appears. As shown in Figure 7, there are two circles with the radius 1, or the length of one side of the square, around the vertices adjoining the main vertex and one circle with the radius $\sqrt{2}$, or the length of diagonal, around the vertex opposite to the main vertex. The region

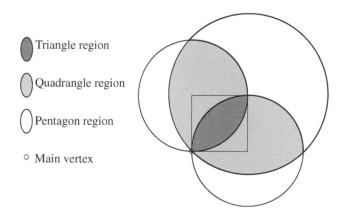

Figure 7. The completed "map" of the TUP regions.

where the three circles overlap is the triangle region, the regions where two circles overlap are the quadrangle regions, and regions where a single circle does not overlap are the pentagon regions.

The first time I taught this exercise, probing the educational use of origamics, I calculated the area of each zone on the recommendation of a mathematics teacher.

Let the side of the square piece of paper have length 1. Then we have the following:

Area of the triangle region $= \frac{\pi}{2} - 1$ (0.5708)   6.89%
Area of quadrangle region $= \pi$ (3.14159)   37.93%
Area of pentagon region $= \frac{\pi}{2} + 3$ (4.5708)   55.18%
Total area of the three regions $= 2\pi + 2$ (8.2832)   100.00%

When I ask students or participants in a class to fold origami once at random, the majority of them produce a triangle and only a few make a pentagon. When we calculate the area of each zone, however, the triangle zone is smallest, or only 6.89% of the total area. In contrast, the pentagon area is largest, 55.18% or more than half of the total area. Why? When we have an origami in hand, we tend to fold it into half or less of the original size, possibly from origami folding experience or simply due to ease in folding.

This subject brings forth further questions, including difficult ones. I will just present some of them here. If you are interested, please try to solve them.

1. Fold a square sheet of paper once and look at the whole shape produced, not only the TUP. What polygon areas will it make? From such a viewpoint, J and K in Figure 1 constitute a hexagon and a nonagon, respectively. Construct a map of regions of movement of the main vertex by type of polygons. Before we had regions of positive area for figures, but here we may have only a point or a line instead. This gives rise to difficulty in drawing the figures and calculating them.

2. Try the same exercise with a rectanglar piece of paper, especially with ratio $1:\sqrt{2}$.

3. Try various regular polygons, such as a regular pentagon, regular hexagon, and so on as the original sheet of paper.

The simple action of folding a square sheet of paper only once has led to an unexpectedly complex, profound subject and produced clear figures. In Japan, we can buy commercial pieces of origami paper (square sheets) cheaply. In the act of folding them for the purposes of mathematics, you must have found quite a different world than when folding an animal or other figure, even though the act of folding remains the same.

The subject I have discussed produced no origami product, no beautiful or cute object. The only physical solid we have are some pieces of square paper with randomly folded lines. People will be puzzled if we show them those pieces of origami. Instead, we have figures of areas constructed using a ruler and compasses. In this project, square sheets of paper were used not as material for art, but as expendables for origamics experiments.

When I started enjoying this subject six years ago, I was a professor of biology studying insect phylogeny. I made an enlarged picture of Figure 7 and pinned it on the wall of my office in the upside-down position. A student came in my office and looked at the figure. He asked me what the figure expressed. I had a sudden fancy to reply, "Have you ever seen the front view of an insect face?" I only asked him that, but he seemed satisfied immediately. He said, "Oh, it has been reduced to geometric lines, but I'm sure it must be a small insect's face like a thrips!" He knew I was studying thrips, insects of the order Thysanoptera during those days, but supposedly he had never seen a front view of a thrips. Now I guess that he probably still believes the figure is an insect face.

2   Haga's First Theorem and Its Application to a Rectangular Piece of Paper

Let's proceed to the second topic, the mathematics of Haga's First Theorem and its generalization. This finding of mine was first described by Dr. Koji Fushimi, former Chair of the Science Council of Japan. (He was also the president of the second *OSME* meeting.) He introduced my finding as "Haga's Theorem" in the January 1979 issue of the monthly journal *Mathematics Seminar*. Now it is called Haga's First Theorem because I later found other theorems.

We start with a square sheet of paper, which I'll call an origami from now on. When I give a colorful set of origami to school students or grown-ups, they soon become eager to fold the paper into an animal, a flower, or other figures. I can direct that impatience to the start of an experiment. I say, "Fold an origami once or twice as you usually do when you fold it into an animal or a flower." When they have started folding, I ask them how they have folded it.

The types of initial folds found in reproducible origami models are things like putting the opposite sides together, or "book fold," and putting the diagonal vertices together, or "triangle fold." If we call this mathematically "folding at a median line" and "folding at a diagonal line", it seems more suitable for origamics.

By asking students how they initially folded an origami, I have them recognize that these two types of folding are the only reproducible ways of folding, and have them think about the reason for this.

What made me find my "mountain of treasures" of origamics was just this simple question: "Why origami folding starts with either of these two ways of folding. Are there other ways?"

A square origami has four sides and four vertices. When we fold an origami by moving a side, we tend to bring it onto a landmark or baseline. At the start, only other sides can play this role. When we fold an origami by moving a vertex, other vertices can be the landmark point. Whichever side or vertex we may use as a base, the resultant fold is either "folding at a median line" or "folding at a diagonal line."

So, I was wondering what will happen if we put a single mark somewhere other than the four sides and four vertices: This led to Haga's First Theorem. The easiest way to put a mark, a reproducible one, on origami without using any tools is to mark the midpoint of a side. As shown in Figure 8, part A, we can mark the midpoint of the top side by bringing together both its ends and making a very small fold or scratch with one's nails. The mark is called a "scratch mark" or "mark".

This single mark enables four new ways of folding. But I'll only describe one of them and leave the others for your own pleasure. We can bring the bottom-right corner (or equivalently the bottom-left corner) onto the scratch mark of the midpoint as in Figure 8, part B. This resultant asymmetric figure has many interesting qualities.

First, let's calculate the side AF of $\triangle AEF$ (Figure 8, part C). Let the length of the side of square be 1, and the length in question be a. The length of three sides of $\triangle AFE$ are $AE = 1/2$, $AF = a$, $EF = 1 - a$. By the Pythagorean Theorem, $(1 - a)^2 = a^2 + (1/2)^2$, so $a = 3/8$, hence, $EF = 1 - a = 5/8$. This means that the fold divided the right side into the ratio of 3 to 5. These values show that the three sides of $\triangle AEF$ are in the ratio

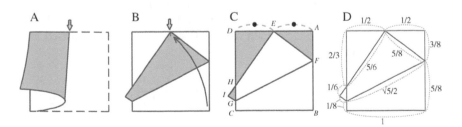

Figure 8. Fold the bottom-right corner to the midpoint of the top edge.

of $3/8:1/2:5/8 = 3:4:5$. Thus the triangle is an "Egyptian triangle" which was used for the measurement of land in the Nile River basin several thousand years ago. When ancient people redrew boundary lines after yearly floods, they stretched a rope knotted at regular intervals into a triangle in the ratio 3:4:5 to make a right angle. If we are to construct an Egyptian triangle, which could be called the starting point of geometry, with a ruler and compass, it requires many procedures and much time. But we can make one by folding a square piece of paper only once after marking the midpoint of a side.

Second, this folding provides us with more interesting observations. Look at $\triangle DEH$ in Figure 8, part C. $\angle DEH$ of the triangle and $\angle AEF$ of $\triangle AEF$ are complementary angles because the right angle B of the square lies between them. Thus $\triangle DEH$ and $\triangle AEF$ are similar right triangles. Now we can calculate the length of the DH. By similar triangles we have the ratio of sides $AF:AE = DE:DH$. That is to say, $3/8:1/2 = 1/2:DH$, so $DH = 2/3$.

This value amazed me, because it shows that we can easily trisect the left side CD by using the point H. People usually trisect a side of origami by sliding the paper little by little until they can fold it in thirds. If we were to do it more mathematically, we would make a rectangle whose ratio of longer side to shorter side is 4:3, and proportionally divide a side using its diagonal. But it is troublesome and leaves useless folds on the paper. Instead, we can trisect a side just by folding a piece of paper only once using the midpoint of a side. By making pinches, it is possible to mark the point H while scarcely leaving any unwanted folds.

The simple action of folding a square piece of paper by putting the bottom-right vertex on the midpoint of the top side divides sides of the square in various ways. We can calculate lengths of all the divided sides in Figure 8, part C based on above-mentioned lengths as indicated in Figure 8, part D.

In what we did above, we marked the midpoint of the top side. What if we mark a point elsewhere on the top side? Let's fold the bottom-right vertex on an arbitrary point as shown by the arrow in Figure 9, and let the point be E and the length of AE be x as also shown in Figure 9. This single folding also produces sections of various length, y_1 through y_6.

Assuming the square has side length 1, we have $AE = x$, $DE = 1 - x$, $AF = y_1$, $BF = 1 - y_1$. So each of the 10 sections can be expressed as a function of x. Namely,

$$y_1 = \frac{(1+x)(1-x)}{2}, \quad y_2 = \frac{2x}{1+x}, \quad y_3 = \frac{1+x^2}{1+x}, \quad y_4 = \frac{(1-x)^2}{2}$$

$$y_5 = 1 - \left(\frac{2x}{1+x} + \frac{(1-x)^2}{2}\right), \quad y_6 = \sqrt{x^2 + 1}.$$

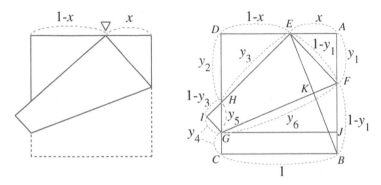

Figure 9. Folding the bottom-right corner to an arbitrary point on the top side.

Such general expressions cannot really convey the wonder, amusement, and utility of this experiment. Tables 1 and 2 show how we can divide the sides of a square with a single piece of paper without using any tools. Initially, we marked the midpoint of a side. We can also easily mark a quarter point, automatically producing the length of three quarters. When we do this, we obtain other interesting lengths for $y_1, ...y_4$, which are shown in Table 1.

The one-digit denominators in Table 1 include 2, 3, 4, 5, 6, 7, and 8. If we fold the length $6/7$ in half, for example, we get $3/7$ and at the same time the other part $4/7$. Thus we can get all fractions of integral numerators over 2 through 8 by this simple process. And by marking the new division points and folding the paper onto them, we can make further and further divisions (see Table 2). In addition, we don't need to crease the paper completely to mark a point on a side. Thus I am flattering myself that this method is highly useful.

With regard to the method of folding either corner at the bottom onto a point on the top side, Mr. Kunihiko Kasahara, an origami researcher, once called it Haga's Theorem in my honor. However, I prefer to call it "the first folding theorem" because I have found some other unique foldings since then.

We can also apply the first folding theorem to a standard $1:\sqrt{2}$ rectangular piece of paper.

x		1/2		1/4		3/4	
y_1	$1 - y_1$	3/8	5/8	15/32	17/32	7/32	25/32
y_2	$1 - y_2$	2/3	1/3	2/5	3/5	6/7	1/7
y_3	$1 - y_3$	5/6	1/6	17/20	3/20	25/28	3/28
y_4	$1 - y_4$	1/8	7/8	9/32	23/32	1/32	31/32

Table 1. Length ratios from the first theorem using easily marked points.

x	y_1	y_2	y_3	y_4	x	y_1	y_2	y_3	y_4
1/3	4/9	1/2	5/6	2/9	1/8	63/128	2/9	65/72	49/128
2/3	5/18	4/5	13/15	1/18	3/8	55/128	6/11	73/88	25/128
1/5	12/25	1/3	13/15	8/25	5/8	39/128	10/13	89/104	9/128
2/5	21/50	4/7	29/35	9/50	7/8	15/128	14/15	113/120	1/128
3/5	8/25	3/4	17/20	2/25	1/9	40/81	1/5	41/45	32/81
4/5	9/50	8/9	41/45	1/50	2/9	77/162	4/11	85/99	49/162
1/6	35/72	2/7	37/42	25/72	4/9	65/162	8/13	97/117	25/162
5/6	11/72	10/11	61/68	1/72	5/9	28/81	5/7	53/63	8/81
1/7	24/49	1/4	25/28	18/49	7/9	16/81	7/8	65/72	2/81
2/7	45/98	4/9	53/63	25/98	8/9	17/162	16/17	148/153	1/162
3/7	20/49	3/5	29/35	8/49					
4/7	33/98	8/11	65/77	9/98					
5/7	12/49	5/6	37/42	2/49					
6/7	13/98	12/13	85/91	1/98					

Table 2. Length ratios from the first theorem using previously marked points.

Do you know about the A-series and B-series sizes of rectangular paper? Japan and many European countries have industrial standards for shape and size of paper of those series, such as JIS of Japan and DIN of Germany. They have the standard ratio rectangle: two sides with ratio $1:\sqrt{2}$. This ratio is convenient because when such paper is cut in half, the resulting pieces are similar, having the same ratio of $1:\sqrt{2}$.

In Japan, most official documents are A4 size. However, this A4 size and other A-series or B-series (A5, A3, B5, B4, etc.) are regrettably not so popular in North America. The most popular here in North America is the 8.5 x 11" size, which is shorter than A4, but a little wider. A- and B-series sizes are similar, the same ratio, but in different size scales. The area of the largest sheet named A0 is 1 square meter while that of B0 is 1.5 square meters.

Let's explore the first folding theorem to $1 \times \sqrt{2}$-sized paper. Place a sheet on the desk horizontally (Figure 10, part A). Mark the midpoint of both top and bottom sides, or fold the paper in half and make a crease of the vertical median (Figure 10, parts B and C). Move the bottom-right (or left) vertex onto the midpoint of the top side and crease the paper (Figure 10, part D). Mark the new position of the midpoint of the bottom side (or the lower end of the vertical median) on the main sheet. Then we get a peculiar mark. How is it peculiar? The mark indicates a mathematically accurate trisection point of the vertical and horizontal sizes! The proof is left as an exercise.

Next, place a new $1 \times \sqrt{2}$ rectangle on the desk in a vertical position (Figure 11, part A). Mark the midpoint of the top side, and fold the paper by the first folding theorem. In the same manner as in a square sheet, we can find several interesting lengths in the folded paper. The longer segment on the right side is

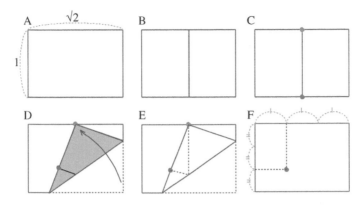

Figure 10. An exercise with $1 \times \sqrt{2}$-sized paper.

divided by the crease into the ratio of 7:9 (Figure 11, part C, AF:BF), while the opposite side is divided by the crease into the ratio of 11:5 (DG:CG). The point H may be more interesting. The ratio of DH:CH is 2/7:5/7. The sheet can be divided into seven equal parts by this point H.

The A-series and B-series sized paper is called "the silver ratio size" or "silver rectangle," compared with the golden ratio of 1:$(1 + \sqrt{5})/2$. I find silver rectangles to be a fascinating paper size to study, along with squares.

3   An Arbitrarily Made Mother Line Bears 11 Wonder Babies

Now let's experience the last wonder. Please take another square piece of paper and fold it once however you like. Figure 11 shows an example where the crease is a line connecting two adjacent sides, but a crease connecting two opposite sides is also acceptable. You could just draw a line using a ruler, of course, but you can get a line simply by folding.

Let's call this crease line the mother line, or generator. I recommend color-marking the mother to distinguish it from other lines to be produced later. We will produce the new lines by folding the paper, bringing each side, or part of the side of the square in line with the mother line. I call such folding "side-onto-line" folding, and it is one of the technical terms of origamics.

We can perform this side-onto-line folding by doing the following: Bring the left part divided by the mother line of the top side in line with the mother line, and fold the paper, making a new crease line. Unfold and flatten the paper, and then bring the bottom side in line with the mother. Since two of

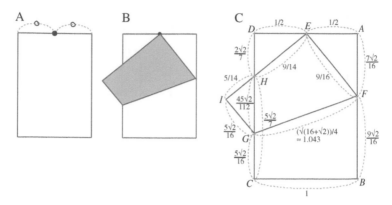

Figure 11. The first folding theorem produces lengths in a $1 \times \sqrt{2}$-sized paper.

the four sides of the square are divided by the mother line, the circumference of the square consists of six parts.

   Please fold all those six parts in the same manner to the mother line. Therefore when you fold the paper, bringing each part in line with the mother, it makes six lines as in Figure 12, part F. Note, if one or both of the ends of your mother line lies in a corner or two corners of the square, then you will have four or five lines. Since the mother was drawn randomly, each of you should have a unique product of creases.

   When you finish making the six lines, put aside the first paper and get a second piece of paper. In contrast to the first paper, where you folded randomly, fold the second paper as follows: Please make the four primary creases the usual way an origami is folded. We first make two rectangles or two triangles by folding the sides together, that is "side-onto-side" folding.

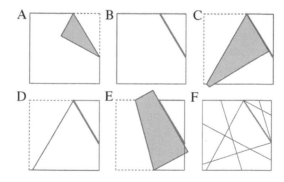

Figure 12. Folding an arbitary mother line and folding all sides to it.

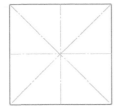

Figure 13. The second paper with primary crease lines compared to the first.

Folding opposite sides together will make a horizontal and a vertical crease, and folding adjacent sides together will fold the diagonals. I call each of these four lines the primary crease, or the primary origami lines.

Let's compare these two sheets: the arbitrary one and the regular one. Is there any relationship between the side-onto-line folds based on an arbitrary line and ones made by standard ways? Apparently, there is no relationship. Now then, let's prominently mark the intersection points of creases in the first sheet, say, with a colored or dark pencil as in Figure 13. Please, count the number of intersections. How many intersections are there on your sheet? Mark and count all cross points between newly made creases, not the mother line and square sides. The number of points varies from two to seven, but usually six or seven.

Next, lay the two sheets one on the other overlapped completely and look through the two sheets against a light (Figure 14). Can you find anything concerning the intersection positions? Don't you see that all the points lie on the four primary lines? You may not see that well through opaque paper. In that case, you can fold the first sheet itself further to make primary folds. Paper allows for this method, although too many lines may be confusing later on.

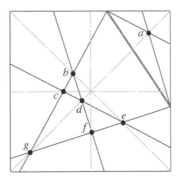

Figure 14. Look! All points are on primary creases!

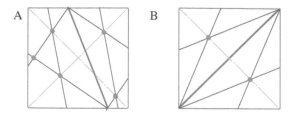

Figure 15. Other examples of mother lines.

Some of you may say, "Yes, all of the points are on primary lines, but some of them are slightly deviated from the line." In a scientific experiment, we should not advocate a law on the basis of a single outcome. We should try a new one using a mother line of different position, length, and direction as in Figure 16, part A. You may even try using the diagonal line as the mother as in Figure 16, part B or the median line of the square.

After several experiments, I came to the conclusion, "All of the intersection points of creases, made by side-onto-line folding based on an arbitrary mother line, lie on primary lines." But it's nothing more than a conjecture. We have to prove that it is always true.

We had seven intersection points in the first square sheet we folded. Let's mark the points with symbols a through g as in Figure 14. We will need several mathematical principles to prove the location of seven points on the primary lines.

Principle 1, Incenter: Let's start with the point a. As shown in Figure 16, point a is in a corner of the square above the mother line. The triangle ABC here is thus a right triangle. Since the two lines which cross each other

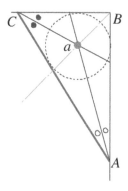

Figure 16. Point a is the incenter of a triangle.

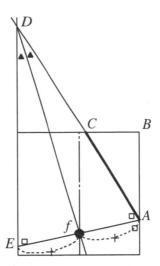

Figure 17. f is the midpoint of the base of an isosceles triangle.

at point a were made by dividing each corner of the triangle in half, they are bisectors of angles A and C. Probably you have learned at school that bisectors of internal angles of a triangle intersect at one point, which conforms to the center of an inscribed circle. Point a is the intersection of bisectors of two internal angles and is precisely the incenter. Therefore, the bisector of the remaining corner should pass through the point a. Well now, this third angle is a corner of the square and is thus a right angle. Its bisector is a diagonal of the square which is just one of the primary creases.

There is another incenter among the remaining points. If we extend the lines of the mother line, the left-hand side and the bottom side, they make a large right triangle, and its incenter is point d.

Principle 2, Midpoint of the base of isosceles triangle: Next, look at point f. Extension of the mother line, the left-hand side, and one of the creases passing point f makes $\triangle ADE$, as shown in Figure 17. E stands for the point where the crease from A passing throught f meets the left-hand side. The crease AE is the bisector of the external angle of A. On the other hand, left and right sides of a square are parallel and thus the alternate interior angles in terms of the crease AE are equal to each other. Thus $\angle DAE$ equals $\angle DEA$. Therefore, $\triangle ADE$ turns out to be an isosceles triangle. Since point f is the intersection of the bisector and bottom of this triangle, Af equals fE, which places it on a primary line. Likewise, point c is the midpoint of an isosceles triangle's side, made by extending the bottom side, the mother line, and line bc.

Principle 3, Excenter: Points b and e can be explained by almost the same principles. Among the five centroids of a triangle, the excenter might be most unfamiliar to you. It is defined as, "the intersection of the bisectors of one internal angle and the external angles of the other two angles of any triangle. It is the center of described circle." As with the other four centroids, every triangle has an excenter. Please look at the isosceles triangle ADE in Figure 17. When we pay attention to the right triangle sticking out above the square, the fold from C through b is the bisector of one external angle of the right triangle. Since the intersection of that fold and the bisector of the vertical angle (like Df) of the triangle is point b, the bisector of the other external angle will also pass through b. But this second external angle is one of primary creases and a diagonal of the initial square. You will see that point e is likewise an excenter if you extend the mother line downwards and extend the bottom side rightwards to see the resultant right triangle poking out of the square.

Now we only have to handle point g. In mathematical classes or courses, students find the most difficulty in proving that this point g is on a primary crease. But the fact is, this point is also the intersection of bisectors of external angles of a triangle like point b and e. Look back at the $\triangle ABC$ we used in the first proof (Figure 16). The bisector of its internal angle B is a diagonal of the square and a primary crease which passes through point g, which is the excenter of this triangle. Incidentally, it is interesting that the described circle round the excenter g is so large that the original triangle would look as if it were parasitic to the circle.

Alternate Proofs: We can locate points b, e, and g without using the theory of excenter. Instead, drop perpendiculars from these points and pay attention to congruence of the resultant right triangles. As an example, let's look at point g in Figure 18. Drop perpendiculars from g to the top side, right side, and the

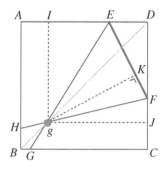

Figure 18. Another way of proving g is on a primary crease.

mother line, and let the feet be I, J, and K, respectively as shown in Figure 18. The right triangles FgJ and FgK are congruent with each other, because $\angle gFJ$ equals $\angle gFK$ and they share the side Fg. Therefore, $Jg = Kg$. The right triangles EgI and EgK are also congruent because $\angle gEI$ equals $\angle gEK$ and the side Eg is common. Therefore, $Ig = Kg$. Hence $Ig = Jg$. Therefore, the quadrangle $IgJD$ is a square, and its diagonal overlaps with the diagonal BD of square $ABCD$. Thus we have that the intersection g is on a primary crease.

Now we discuss the mystery of intersection points on crease line extensions. I told you previously that the number of intersections of the creases ranges from two to seven depending on the position and length of the mother line. But even when there are seven intersection points as in Figure 16, the number is reduced to six if the mother line is constructed a little longer as in Figure 19, part A. Of course, all of these six points are on the primary crease. Now then, why should such variation occur? It is because point g in this figure has jutted out of the paper. If we extend the two folds running toward the lower left, their intersection point g appears outside the square. We can prove that the point lies also on the extension of a primary crease by the theory of excenter or congruence of right triangles previously discussed.

When I discovered this phenomenon, I perceived that extension of another crease might make further intersections outside the initial paper. So I extended

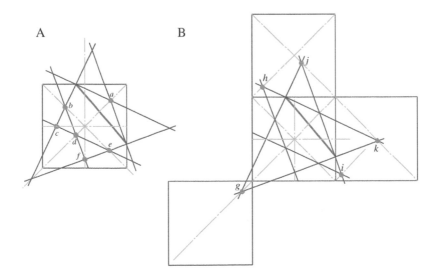

Figure 19. More intersections are found by extending the crease lines.

all promising lines and got intersections as Figure 19, part B shows, that is, points h, i, j, and k. I expected in high spirits that those points, too, would be on the primary lines. But, to my disappointment, the points apparently deviated from the extension of the primary crease lines. However, when I casually slid the paper with primary fold lines to the right just by its width, the points i and k lay perfectly on the lines. How mysterious! I then slid the paper with primary fold lines upwards just by its height. The points h and j were on the lines as well. I was very much encouraged and constructed different mother lines in many other pieces of paper, producing intersections outside the paper by extension, and slid the paper with primary fold lines to every direction. Every point lay on primary lines without exception. Since we can say that the first point g produced by extension is also on a primary fold line that was slid off the lower left, we get the conclusion, "Intersections which appear by extension of folds made by side-onto-line folding based on an arbitrary mother line of any length and inclination lie on primary fold lines in a paper slid horizontally, vertically, or diagonally just by its size."

In order to solve this puzzle, I'd like to give new names to some points except for the intersections. The intersections h and i can be solved by the same principles. In Figure 20, part A, the extended mother line crosses the extensions of the left side of the initial square and of the bottom side at G and H, respectively. The fold produced by folding the square bringing the left side AB in line with the mother line and its extension is GO, which bisects $\angle AGE$. The line KL is the bisector of $\angle AEG$, or $\angle DEF$. The intersection h of GO and KL, which are both bisectors of internal angles of $\triangle AEG$, is the incenter of that triangle. Thus the bisector of its last angle should also pass

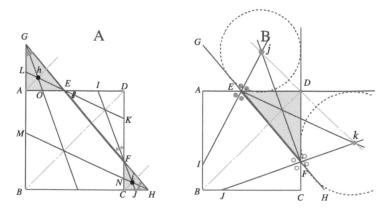

Figure 20. Intersections h and i are incenters, while j and k are excenters.

the incenter. The angle in question is a right angle, and its bisector is one of primary lines, or a diagonal, of the paper slid upwards by its height. Likewise, the fold made by having the bottom line meet the mother line and its extension MH is the bisector of $\angle FHC$ and the line IJ is the bisector of $\angle CFH$ or $\angle DFE$. The intersection i of MH and IJ which are bisectors of two internal angles of $\triangle CFH$ is thus the incenter of the triangle. The bisector of the last angle should pass the incenter, and that bisector is a diagonal fold in the paper slid rightwards just by its width.

Let's examine next points j and k. (See Figure 20, part B.) Fold EI is the bisector of $\angle AEF$. Its extension Ej is thus the bisector of $\angle DEG$, which is vertically opposite to $\angle AEF$. $\angle DEG$ is one of internal angles of $\triangle DEF$, and the fold and its extension Fj is the bisector of $\angle DFE$, which is one of internal angles of $\triangle DEF$. Thus the intersection j of Ej and Fj is an excenter of $\triangle DEF$. Therefore, the bisector of the right angle to the upper left of D, which is another external angle, should also pass through excenter j, which is a diagonal primary fold in the paper slid upwards by its height. Likewise, fold FJ is the bisector of $\angle CFE$ and its extension Fk is the bisector of $\angle DFH$ that is vertically opposite to $\angle CFE$. And this $\angle DFH$ is one of external angles of $\triangle DEF$. Since the fold and its extension Ek is the bisector of $\angle DEF$ that is one of internal angles of $\triangle DEF$, the intersection k of Fk and Ek is another excenter of $\triangle DEF$. Therefore, the bisector of another external angle, namely the right angle to the lower right of D should pass excenter k and the bisector is a diagonal primary fold in the paper slid rightwards by its width.

An arbitrary mother line produced six creases in a square sheet of paper by bringing its sides in line with the mother. They produce 11 intersections, including those on the extensions of the creases, and all of them were found to lie on primary fold lines. This phenomenon was initially given the romantic name "Seven Stars" because a maximum of seven intersections appeared in the original square. Later, intersections on the extensions appeared additionally, so the phenomenon can be called "Wonder Eleven Stars," now, instead of "Seven Stars."

Several years ago, I unintentionally found this mysterious phenomenon in folding origami, and at the time, I could hardly believe it. I remember that I folded one origami after another changing the position of the mother line. I was afraid that a quick solution of the principles would spoil my pleasure, and folded and folded origami for sometime. I did this partly because I anticipated further evolution of this experiment.

When we encounter such mysterious phenomena, we become inclined to share its proofs with others. We don't want to say who the criminal is before others read a mystery novel we recommend, but at the same time, we are impatient for others to share the fun. Origamics gives me the pleasures of discovery, of scientific solution, of mathematical expansion, and of educational practice. In addition, it gives me the pleasure of passing this on to great audiences like the readers of this paper.

I express my hearty thanks to the organizer, Professor Thomas Hull, and OrigamiUSA for their kind invitation and arrangement of such a wonderful meeting and this proceedings.

Origami as a Model for Development in Organisms

Norman Budnitz

1 Introduction

With the advent of modern genetics, some humans have been intrigued by the notion that if we could control our genetic makeup, we could then control such things as inherited diseases and congenital abnormalities. Others have expressed concern that this same genetic control might allow us to choose particular traits in our offspring, thus interfering with natural processes. In some sense, both of these ideas are correct, but are based on a very limited understanding of how our genes actually influence our physical and behavioral characteristics.

Governments, universities, and private industries have poured enormous amounts of money into the Human Genome Project, a program aimed at describing the sequence of nucleotides in our DNA. Since genes are made up of nucleotides, the idea is that once we know the overall sequence, we will be well on our way to knowing where the genes are. Once we know where the genes are, we can determine what they do. Once we know what they do, we can get rid of the bad ones and enhance the good ones. The problem is that the developmental process from fertilized egg, through embryo, fetus, childhood, adolescence, to eventual adulthood is influenced by a multitude of factors that go well beyond the initial schematic diagram, the genes.

Some genes do directly code for such things as hair color or the ability to digest milk products. But other genes control the actions of these first genes, turning them on or off, or controlling the rate at which they do their jobs. Still other genes determine larger developmental issues, such as the general layout of an organism—head, trunk, limbs—or the sequence of events—limb buds, arms, fingers. In addition to these complex gene interactions, there are environmental influences. If your parents are tall, you are likely to be tall. But if your mother feeds you well during your growing years, you will likely be even taller. And if your mother feeds *herself* well while you are *in utero*, you will probably be taller still.

There are still other influences beyond genes that effect development. As an organism develops, the growing and changing tissues produce chemicals that affect the growth and development of neighboring tissues. Hormones from a mammalian mother effect changes in the embryo (and vice versa). We now have evidence that chemicals in the environment that mimic hormones can also cause changes in embryonic development.

The point of this exercise is to help students get a feel for the complexity of these developmental processes.

2   Procedure

Morphogenesis is the process by which one gets from gene to protein to extra-cellular structure to embryo. In a playful and illustrative example a flat square of paper can become a swan via the process of origami. Figure 1 shows the origami *crease pattern* for making a swan. In this diagram, dotted lines are *valley folds* (the crease forms a valley when seen from this side of the paper), and solid lines are *mountain folds*. Students can cut out the square and then try to fold the swan just knowing this much information. Give them five to ten minutes to try this, but since it is very difficult to accomplish, do not let them get frustrated.

After the students have struggled a bit with the crease pattern, hand out squares of paper and lead them through the process as described in, the step-by-step diagrams of Figure 2. Eight-and-a-half-inch squares cut from copy paper work just fine, or you can supply 10-inch origami paper that you can buy in most hobby stores. When they are finished and have exclaimed over the beauty of their swans, have them unfold their models and compare the crease pattern to Figure 1.

The *pattern* of folds is the *developmental program* of the origami swan. All the necessary folds are shown, including their direction (mountain or val-ley). But the *order* of folds is also important. It is much easier to make the long diagonal folds first (and in a particular order) and then make the little neck and head folds (the arcs of mountain and valley folds) all at once. The

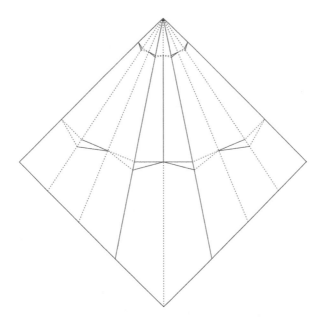

Figure 1. Crease pattern for a swan.

developmental program of a *real* swan is somehow contained in the genes. It is fair to say that we do not know the complete developmental program for this complex organism. In fact, even in our model, there are some confusing "instructions". Note that although the arcs of mountains and valleys come in pairs, only one of each of these folds occurs in the finished swan. One member of each pair is formed during the folding process (Steps 5 and 6), but is superseded when the head and neck are pulled up (Step 8). This is one reason folding the swan directly from the crease pattern (Figure 1) is difficult. We decided to include both members of each pair in the crease diagram because this is what the students will see when they unfold their models.

Figure 3 shows the crease pattern for a duck. Though there are similarities to the swan crease pattern (long diagonals and arcs of mountain and valley folds), there are also distinct differences (*fewer* diagonals and *more* arcs, for example). Point out these differences to students and ask them to predict how they might affect the final origami model. Figure 4 gives the step-by-step instructions for the duck. After making the duck, students should be able to identify the differences. The fewer number of long diagonals leads to a "fatter" bird. The extra arcs appear in the head and beak shape. And the turned up tail is a new feature altogether.

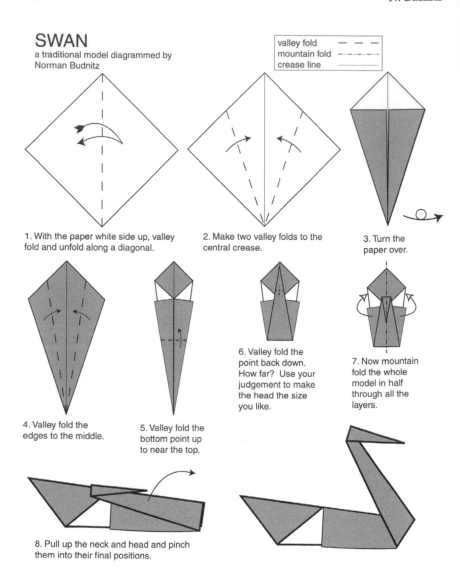

SWAN
a traditional model diagrammed by
Norman Budnitz

valley fold	– – –
mountain fold	–··–··–··
crease line	——

1. With the paper white side up, valley fold and unfold along a diagonal.

2. Make two valley folds to the central crease.

3. Turn the paper over.

4. Valley fold the edges to the middle.

5. Valley fold the bottom point up to near the top.

6. Valley fold the point back down. How far? Use your judgement to make the head the size you like.

7. Now mountain fold the whole model in half through all the layers.

8. Pull up the neck and head and pinch them into their final positions.

Figure 2. Instructions for folding a swan.

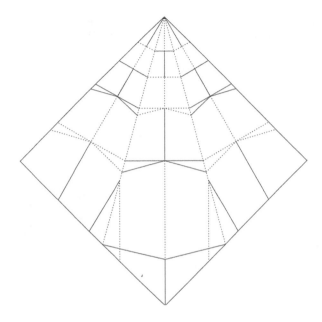

Figure 3. Crease pattern for a duck.

3 Summary

The expression of genes as structure (genotype to phenotype) is a complex process. This origami exercise is an analogue to the real-world biological series of events.

1. The *crease pattern* is a model of the genome. The folds are all shown but are insufficient information for completing the model.

2. The "extra" folds—the initial folds of each arc that do not actually occur in the finished model—are analogous to developmental genes that are important early in life, but that do not get used later on. For example, in humans, there is a gene for a fetal form of hemoglobin that is superseded by a gene for an adult form of hemoglobin. The fetal hemoglobin works better *in utero*, while the adult form works better when the infant starts breathing air on its own.

3. The *step-by-step instructions* are analogous to the *sequence* "instructions" that translate genetic information into the making of proteins (enzymes) that further direct the growth and development of the living organism.

DUCK
diagrammed by Norman Budnitz

valley fold	– – –
mountain fold	·–··–··–··
crease line	——

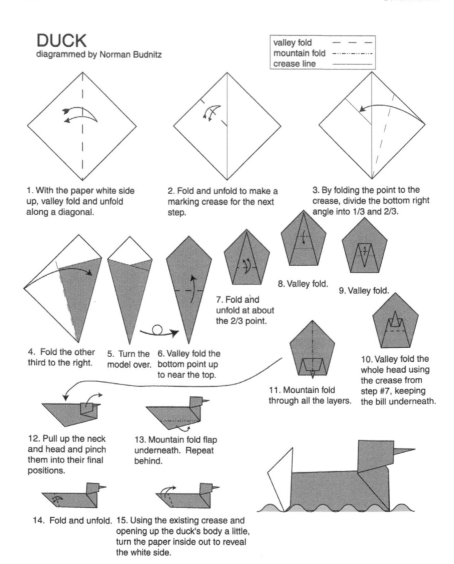

1. With the paper white side up, valley fold and unfold along a diagonal.

2. Fold and unfold to make a marking crease for the next step.

3. By folding the point to the crease, divide the bottom right angle into 1/3 and 2/3.

4. Fold the other third to the right.

5. Turn the model over.

6. Valley fold the bottom point up to near the top.

7. Fold and unfold at about the 2/3 point.

8. Valley fold.

9. Valley fold.

10. Valley fold the whole head using the crease from step #7, keeping the bill underneath.

11. Mountain fold through all the layers.

12. Pull up the neck and head and pinch them into their final positions.

13. Mountain fold flap underneath. Repeat behind.

14. Fold and unfold.

15. Using the existing crease and opening up the duck's body a little, turn the paper inside out to reveal the white side.

Figure 4. Instructions for folding a duck.

4. Comparing the crease patterns of the swan and duck is analogous to comparing the genomes of those two organisms. (The genomes of humans and chimpanzees only differ by about 1%. That is, it is estimated that 99% of the genetic information carried in the cells of these two species is identical!)

5. The differences seen in the crease patterns are analogous to *mutations* in the genes of real organisms.

6. Changes in the sequence of folding, sloppy folding, and spilled coffee are analogous to changes that occur when environmental factors cause changes in the development of organisms.

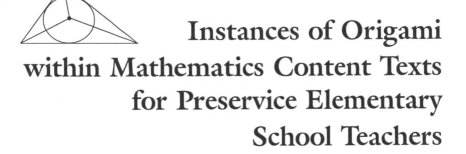

Instances of Origami within Mathematics Content Texts for Preservice Elementary School Teachers

Jack A. Carter and Beverly J. Ferrucci

1 Introduction

Origami, the art of paper-folding, receives substantial endorsement from current reform initiatives in mathematics education. Particularly, the National Council of Teachers of Mathematics in its *Principles and Standards for School Mathematics* recommend that elementary school students use paper-folding to perform their initial investigations of symmetry and other mathematical relationships (NCTM, 2000). Since prospective elementary school teachers need a solid foundation in the mathematical basis of these reform initiatives, mathematics textbooks for these future teachers should contain ample models of paper-folding applications. A preliminary review of the literature showed there were no recent or systematic surveys of the nature and extent of these applications. Consequently, a survey was undertaken to determine the nature and extent of origami in current textbooks intended for use in mathematics content courses for future primary and intermediate school teachers.

2 Procedures

To conduct the survey, a selection of ten popular textbooks for prospective elementary school teachers was used (Bassarear, 2001; Bennett and Nelson,

2001a; Bennett and Nelson, 2001b; Billstein, Libeskind, and Lott, 2001; Long and DeTemple, 2000; Masingila and Lester, 1998; Miller, Heeren, and Hornsby, 2001; Musser, Burger, and Peterson, 2001; O'Daffer, Charles, Cooney, Dossey, and Schielack, 1998; Wheeler and Wheeler, 1998). These texts (cited only by the first author's last name in the rest of this report) were chosen based on an informal poll of the researchers' colleagues who also taught courses in mathematics for prospective teachers. The textbooks were all published from 1998 to 2001 and the latest edition of each book was used in the study.

After the ten textbooks were collected, a panel of two graduate students and two Ph.D.s in mathematics education was used to classify the instances of paper-folding in the textbooks. To complete this classification task, the panel met on two separate days for approximately four hours per day. During each of these sessions, panel members reviewed the textbooks and noted any instances of origami and the nature of those instances. Also during each of the sessions, a consensus was reached about the paper-folding opportunities according to their type and nature. At the end of the two sessions, there were no unresolved differences among the panel members with respect to the extent and nature of the occurrences of origami in the textbooks. Early in the panel's work, it became evident that the texts included activities that involved paper-cutting or tearing as well as paper-folding. Definitions of origami usually stipulate that cutting, gluing, or drawing on the paper is to be avoided, and only paper-folding is used to create the desired result (Gross, 1995). As a result, panel members only classified as origami those paper-folding opportunities that did not involve cutting, gluing, tearing or drawing on paper.

3   Results

Table 1 shows the number of paper-folding activities, examples, exercises, and the total number of paper-folding opportunities that were found in the ten textbooks that were surveyed.

Since activities encouraged the reader to actively participate in learning concepts, it was anticipated that most of the texts would contain instances of paper-folding under this classification. However, Table 1 shows that three texts (Long, Bennett(a), and Masingila) contained three-fourths of the paper-folding activities. About 40% of the paper-folding activities were in the Long textbook, and four of the surveyed texts contained no paper-folding activities as they were defined in this study. Four texts (Long, Bennett(a), Bennett(b), and Masingila) contained nearly 88% of the paper-folding activities, but only the Long textbook also contained examples (2) of paper-folding, while the Bennett(a), Bennett(b), and Long textbooks also contained exercises in paper-

Textbook	Number of Paper-folding Activities	Number of Paper-folding Examples	Number of Paper-folding Exercises	Total
Bassarear	0	0	8	8
Bennett (a)	13	0	1	14
Bennett (b)	9	0	1	10
Billstein	6	7	5	18
Long	29	2	3	34
Masingila	12	0	0	12
Miller	0	0	0	0
Musser	0	8	8	16
O'Daffer	0	1	1	2
Wheeler	3	6	14	23
Total	72	24	41	137

Table 1. Paper-folding opportunities in ten selected mathematics textbooks for prospective elementary school teachers.

folding. Information about examples and exercises related to folding paper is described next.

The number of examples in Billstein (7), Musser (8), and Wheeler (6) accounted for all but three of the paper-folding examples found in the ten textbooks. With respect to paper-folding exercises, there were some contrasts between the Billstein, Musser, and Wheeler texts. Table 1 shows that Billstein followed up examples of folding paper with five exercises in paper-folding, whereas the Musser text contained an equal number of exercises and examples (8), and the Bassarear contained eight exercises but no examples. Also, the Wheeler text contained more than a third of the total number of paper-folding exercises in the ten textbooks.

Table 1 further shows there were 137 instances of paper-folding or paper-folding opportunities in the ten texts surveyed. This produced an overall average of 13.7 and a median value of 13 paper-folding opportunities per textbook. However, one of the texts (Miller) contained no instances of paper-folding, while O'Daffer contained only two paper-folding opportunities. On the other hand, the textbook that contained the most opportunities (Long with 34) contained more than twice the average number of opportunities found in the ten textbooks.

There were also contrasts evident with respect to the nature of the paper-folding opportunities. Table 2 shows that angles, properties of triangles, and three-dimensional figures, which appeared in four of the texts, were the most popular types of paper-folding opportunities in the ten texts.

Textbook	Nature of Opportunities	
Bassarear	symmetry	
Bennett (a)	segments	three-dimensional figures
Bennett (b)	linear measurement properties of triangles	polygons
Billstein	angles linear measurement properties of triangles symmetry	parallel lines perpendicular segments polygons
Long	angle trisection fractions properties of triangles	perpendicular segments polygons three-dimensional figures
Masingila	fractions	
Miller	none	
Musser	angles properties of triangles spatial visualization	perpendicular segments Pythagorean Theorem
O'Daffer	symmetry	three-dimensional figures
Wheeler	angles	three-dimensional figures

Table 2. Nature of paper-folding opportunities.

Additionally, perpendicular segments, polygons, and symmetry each appeared in three of the surveyed textbooks. There were also paper-folding opportunities that rarely appeared in the textbooks. These included fractions and linear measurement that were evident in two books and parallel lines, the Pythagorean Theorem, segments, and spatial visualization which each were found in only one of the textbooks. These rarely occurring paper-folding opportunities occurred in six of the textbooks surveyed: Bennett(a), Bennett(b), Billstein, Long, Masingila, and Musser.

4 Conclusions

Although 10 texts were surveyed in the current study and 137 instances of paper-folding were identified, nearly one quarter or 34 of these opportunities to fold paper were in the Long textbook. The Wheeler text contained the next largest number of paper-folding opportunities (23), and this accounted for approximately 17% of the total number of opportunities. The third most frequent number of paper-folding opportunities was contained in the Billstein text.

Textbook and (Number of Paper-folding Opportunities)	Percentage of Total Number of Paper-folding Opportunities
Long (34)	25%
Wheeler (23)	17%
Billstein (18)	13%
Musser (16)	12%

Table 3. Textbooks containing the largest number of paper-folding opportunities.

These accounted for 18 or 13% of the paper-folding opportunities. The fourth most frequent number of paper-folding opportunities (16 or 12%) was in the Musser text. Thus, the four textbooks (Long, Wheeler, Billstein, and Musser) that contained the largest number of paper-folding opportunities, collectively accounted for more than two-thirds of the total number of paper-folding opportunities in the ten textbooks. Table 3 summarizes this information for the texts with the most paper-folding opportunities.

With respect to the nature of the paper-folding opportunities, the most popular topics were angles, properties of triangles, and three-dimensional figures along with perpendicular segments, polygons, and symmetry. The most rarely appearing topics that were addressed by paper-folding included fractions and linear measurement as well as parallel lines, the Pythagorean theorem, segments, and spatial visualization. Table 4 summarizes the nature of the paper-folding opportunities with respect to their frequencies of occurrence.

Based on the findings of this study, two textbooks (Long and Wheeler) contained 25% and 17%, respectively, of the total number of paper-folding opportunities. These percentages exceeded the expected percentage of paper-folding opportunities (10%) that would be the case in the event that these

Topics of Paper-folding Opportunities	Frequency of Occurrence
Angles	4
Properties of Triangles	4
Three-dimensional Figures	4
Perpendicular Segments	3
Polygons	3
Symmetry	3

Table 4. Nature of paper-folding opportunities by frequency of occurrence.

opportunities were randomly distributed among the ten textbooks. Two other texts (Billstein and Musser) contained 13% and 12%, respectively, of the total number of paper-folding opportunities, thereby providing somewhat less convincing evidence that these texts also exceeded the expected 10% level.

The evidence from this study regarding the nature of the paper-folding opportunities revealed that topics in informal geometry, geometric constructions, and symmetry received coverage in some, but not a majority of the textbooks surveyed. Particularly, between three and four of the texts surveyed, included topics from informal geometry, geometric constructions, and symmetry as opportunities for folding paper. Notable among the topics that received less attention by origami were fractions, linear measures, parallel lines, the Pythagorean Theorem, and segments. Fewer than three of the textbooks contained instances wherein paper-folding was applied to these topics. Given the importance of fractions in school mathematics and the difficulty students frequently associate with the topic, it is remarkable that mathematics textbooks for prospective elementary school teachers contain so few of the newer methods for teaching fractions by means of origami (Camblin, 1998; Sinicrope and Mick, 1992). Recent design and development work in origami has also produced opportunities for folding paper that appropriately address still other elementary school topics: tessellations (Stewart, 1999; Neale and Hull, 1994), investigations with geometric objects and their constructions (Morrow, 2001), and measurement of three-dimensional figures (Higginson and Colgan, 2001).

During the course of reviewing the ten texts for instances of origami, the members of the panel that enumerated and classified the opportunities to fold paper made several spontaneous comments concerning the diverse presentations and lack of an unified approach to origami that was generally evident in the ten texts. These qualitative findings point toward the need for a more systematic treatment of paper folding in mathematics content texts for prospective teachers. Smart (1998) made a valuable contribution toward developing a more systematic treatment by illustrating and describing four basic geometric constructions through paper-folding. These constructions are apt to provide prospective elementary school teachers with an opportunity to catch a more unified spirit of paper-folding constructions and to contrast this spirit with that of other methods.

As research and development in origami continues, future advances are likely to produce educational gains in still other areas. Particularly, wet folding (Wu, 2001) and hypergami or the use of computers to design and construct models for origami (Eisenberg and Nishioka, 1996) both show great promise for empowering teachers who use origami to plant and nourish the seeds of mathematical thinking.

References

[1] Bassarear, T. (2001), *Mathematics for Elementary School Teachers, Second Edition*, Boston: Houghton Mifflin Company.

[2] Bennett, A., and Nelson, T. (2001a), *Mathematics for Elementary Teachers: An Activity Approach, Fifth Edition*, New York: McGraw-Hill Company, Inc.

[3] Bennett, A., and Nelson, T. (2001b), *Mathematics for Elementary Teachers: A Conceptual Approach, Fifth Edition*, New York: McGraw-Hill Company, Inc.

[4] Billstein, R., Libeskind, S., and Lott, J. (2001), *A Problem Solving Approach to Mathematics for Elementary School Teachers, Seventh Edition*, Reading, MA: Addison Wesley Longman, Inc.

[5] Camblin, B.A. (1998), "The Mathematics in Your Note Paper", *Mathematics Teaching in the Middle School*, 4(3), 168–169.

[6] Eisenberg, M., and Nishioka, A. (1996), "Polyhedral Sculpture: The Path from Computational Artifact to Real-World Mathematical Object", *Proceedings of the Annual National Educational Computing Conference (NECC 96)*, Minneapolis, MN, June 12, 1996.

[7] Gross, G.M. (1995), *The Origami Workshop*, New York: Michael Friedman Publishing Group.

[8] Higginson, W. and Colgan, L. (2001), "Algebraic Thinking through Origami", *Mathematics Teaching in the Middle School*, 6(6), 343–349.

[9] Long, C., and DeTemple, D. (2000), *Mathematical Reasoning for Elementary Teachers, Second Edition*, Reading, MA: Addison Wesley Longman, Inc.

[10] Masingila, J., and Lester, F. (1998), *Mathematics for Elementary teachers via problem solving, Preliminary Edition*, Upper Saddle River, NJ: Prentice Hall, Inc.

[11] Miller, C., Heeren, V., and Hornsby, J. (2001), *Mathematical Ideas, Ninth Edition* and *Expanded Ninth Edition*, Reading, MA: Addison Wesley Longman, Inc.

[12] Morrow, J. (2001), Jim Morrow's Investigations in Geometry. Accessed in January 2001: http://www.mtholyoke.edu/courses/jmorrow/info.html.

[13] Musser, G., Burger, W., and Peterson, B. (2001), *Mathematics for Elementary Teachers, Updated Edition*, New York: John Wiley and Sons, Inc.

[14] National Council of Teachers of Mathematics. (2000), *Principles and Standards for School Mathematics*, Reston, VA: The Council.

[15] Neale, R. and Hull, T. (1994), *Origami, Plain and Simple*, New York: St. Martin's Press.

[16] O'Daffer, P., Charles, R., Cooney, T., Dossey, J., and Schielack, J. (1998), *Mathematics for Elementary Teachers*, Reading, MA: Addison Wesley Longman, Inc.

[17] Sinicrope, R. and Mick, H.W. (1992), "Multiplication of Fractions through Paper Folding", *Arithmetic Teacher*, 40(2), 116–121.

[18] Smart, J.R. (1998), *Modern Geometries, Fifth Edition*, Pacific Grove, CA: Brooks/Cole Publishing Company.

[19] Stewart, I. (1999), "Origami Tessellations", *Scientific American*, 280(2), 100–101.

[20] Wheeler, R., and Wheeler, E. (1998), *Modern Mathematics, Tenth Edition*, Dubuque, IA: Kendall/Hunt Publishing Company.

[21] Wu, J. (2001), Joseph Wu's Origami Page. Accessed in January 2001: http://www.origami.as.

Contributors

Roger C. Alperin Department of Mathematics, San Jose State University, San Jose, CA 95192, alperin@math.sjsu.edu

Alex Bateman The Wellcome Trust Sanger Institute, Cambridge, Hinxton, CB10 1SA, UK, agb@sanger.ac.uk

sarah-marie belcastro Depertment of Mathematics and Computer Science, Xavier University, Cincinnati, OH 45207-4441, smbelcas@cs.xu.edu

Ethan Berkove Department of Mathematics, Lafayette College, Easton, PA 18042, berkovee@lafayette.edu

Marshall Bern Palo Alto Research Center, 3333 Coyote Hill Rd., Palo Alto, CA 94304, bern@parc.com

Norman Budnitz Center for Inquiry-Based Learning, Depertment of Biology, Duke University, Box 90338, Durham NC 27708-0338, nbudnitz@duke.edu

Krystyna Burczyk Wojciech Burczyk, ul. Starowiejska 21/6, 32-800 Brzesko, Poland, burczyk@mail.zetosa.com pl

Dr. Jack Carter Depertment of Mathematics & Computer Science, California State University, Hayward, CA 94542-3092, USA, jcarter@csuhayward.edu

Erik D. Demaine MIT Laboratory for Computer Science, 200 Technology Square, Cambridge, MA 02139, USA, edemaine@mit.edu

Martin L. Demaine MIT Laboratory for Computer Science, 200 Technology Square, Cambridge, MA 02139, USA, mdemaine@mit.edu

Brian DiDonna Department of Physics and Astronomy, University of Pennsylvania, 209 S 33rd St., Philadelphia, PA 19104, bdidonna@physics.upenn.edu

Jeff Dumont Killington Mountain School, P.O. Box 509, Killington, VT 05751

David Eppstein Dept. of Information and Computer Science, University of California, Irvine, Irvine, CA 92697-3425, eppstein@ics.uci.edu

Dr. Beverly Ferrucci Mathematics Department, Keene State College, Keene, NH 03435-2001, bferrucc@keene.edu

Deborah Foreman-Takano Doshisha University, Kyotanabe-shi Kyoto 610–394, Japan, dforeman@mail.doshisha.ac.jp

Emma Frigerio Universita di Milano, Departmento di Mathematica, Via C. Saldini, 50-20133 Milano, Italy, Emma.Frigerio@mat.unimi.it

Tomoko Fuse Shionokai Yasaka-mura, Kitaazuni-gun Nagano, 399-7301, Japan, ftomoko@mail.valley.ne.jp

Robert Geretschäger Breinergasse 23, A-8051 Graz, Austria, robert.geretschlaeger@brgkepler.at

Kazuo Haga Haga's Laboratory for Science Education, #514/16-2, Higashiarai, Tsukuba, Ibaraki, 305-0033 Japan, hagak@hi-ho.ne.jp

Barry Hayes Post Box 987, Palo Altho, CA 94302, bhayes@cs.stanford.edu

Lillian Y. Ho City College of San Francisco, John Adams Campus/ESL, 1860 Hayes Street, San Francisco, CA 94117, lillianh@ips.net, ih@ccsf.org

Thomas C. Hull Department of Mathematics, Merrimack College, North Andover, MA 01845

Miyuki Kawamura
☐
Toshikazu Kawasaki Sasebo College of Technology, Okishin 1-1, Sasebo, Nagasaki, 857-1193 Japan, kawasaki@post.cc.sasebo.ac.jp

Eva Knoll 308 chemin du Tour, Ile Bigras, Laval, Quebec, Canada, H7Y 1H2, evaknoll@netscape.net

Biruta Kresling Bionics & Experimental Design, 170, rue Saint-Charles, F-75015 Paris, France, bkbionik@aol.com

Robert J. Lang Origami Art & Engineering, 899 Forest Lane, Alamo, CA 94507, robert@lang.cc

Jun Maekawa MAekawa Co., Ltd., 1-1-21-805 Tobitakyu Chofu, Tokyo 182-0036, Japan, maekawa@nro.nao.ac.jp

Koryo Miru 3-9-7 Tsurukawa, Machida, Tokyo 195-0061, Japan, miurak@gakushikai.jp

Charlene Morrow Director, SummerMath, Mt. Holyoke College, 50 College Street, South Hadley, MA 01075-1441, cmorrow@mtholyoke.edu

Dr. Jeannine Mosely Origami Geometrics, 32 Poplar Street, Belmont, MA, 02478, j9@alum.mit.edu

Akira Nagashima Daiei Dream Co, Ltd., 158-7 Tottori-machi Maebashi, Gunma 371-0131, Japan, nagaaki@daieidream.co.jp

Noriko Nagata 1-7-26 Nagahama, Kanazawa-ku, Yokohama, 236-0011, Japan, k5nagata@plum.ocn.ne.jp

Radhika Nagpal MIT Artificial Intelligence Lab, 545 Technology Square, Rm 432, Cambridge, MA 02139, radhi@ai.mit.edu

Yasuhiro Ohara Kuramae Industries Co., Ltd., 176-44 Kamioshima-cho Maebashi, Gunma 379-2153, Japan, yasuhiro_o@kuramae.co.jp

Hiroshi Okumura Maebashi Institute of Technology, 460-1 Kamisadori Maebashi, Gunma 371-0816, Japan, okumura@maebashi-it.ac.jp

Benedetto Scimemi Universita Degli Studi di Padova, Dipartimento do Mathematica Pura ed Applicata, I-35131 Padova, Italy, scimemi@math.unips.it

John Szinger Zing Man Productions, 63 LaBelle Road, Fleetwood, NY 10552, john@zingman.com

Arnold Tubis Professor Emeritus of Physics, Purdue University, West Lafayette, IN 47907, atu@physics.purdue.edu

Index

affine map, 40
algebraic geometry, 84
algorithm, 3–11, 122–125
Alhazen's problem, 83–93, 116–117
amorphous computing, 220
angle deficit theorem, 290
Archimedian solids, 177, 275–277
artificial life, 219, 220
atlas design, 137–144
axioms, origami, 83, 91, 107, 220, 222–227

bamboo, 197
base
    bird, 61–73
    generalized bird, 68–69
base, uniaxial, 5, 10
biology, 228
    developmental, 219–221, 333
    education, 333
biomimetics, 198, 205

bionics, 198, 201
bird base, 61–73
boxes
    isosceles triangular from rectangles, 303–306
    maximum volume and face area, 305–306
    twist, 179–186
buckling, 192–194, 201, 205

calculus education, 303–306
cellular automata, 219, 220
chessboard, 7
circular origami, 285
closed path, 41, 47
colorings, 271–283
complete quadrilateral, 90, 91
complexity, 4, 5, 11
computational complexity, 3–13
computational geometry, 7
computational intractability, 3
computational origami, 3–13
computer simulation, 126, 129–135
cone, 34, 36, 77–78, 148, 149, 198